WINERY TECHNOLOGY & OPERATIONS

A Handbook for Small Wineries

to Ralph

Dr. Yair Margalit

THE WINE APPRECIATION GUILD
San Francisco

ISBN 0-932664-66-0

Requests for permission should be addressed to:

THE WINE APPRECIATION GUILD LTD.
155 Connecticut Street
San Francisco, CA 94107
FAX: 415-864-0377

Library of Congress Cataloging-in-Publication Data

Margalit, Yair, 1938 -
    Winery technology & operations.

    1. Wine and wine making. I. Title.
TP548.M335 1990   663'.2   90-12342
ISBN 0-932664-66-0

Printed in the United States of America
10 9 8 7 6 5 4 3 2 1

Managing Editor: Maurice T. Sullivan

Editorial Assistant: Shannon Essa

Technical Consultants: Robert Kozlowski, Ph.D.
                          Lenny L. Replogle, Ph.D.

Preface

It has been said that winemaking is the second oldest profession. Whether this is true or not, we need not argue. Winemaking is certainly a very old operation and for many years, often poorly done. Only since Pasteur's time, in the mid-nineteenth century, have people begun to understand the science of winemaking. As a result of the developments in chemistry and microbiology and specific research in this field, the understanding and consequent development of its technology have improved rapidly in recent decades.

Many excellent books on winemaking have been published. By writing this book, I felt that I would fill an "empty space" in winemaking literature; a book which is solely oriented toward the practical everyday work of the winemaker, from harvesting the grapes through bottling and laboratory analysis.

This book is called "A Handbook for Small Wineries" because the vast majority of wineries around the world are small. Although their total wine production is also small compared with the total production of the big wineries, small wineries lead in individuality and quality in their wines.

The basic principles of small winery operation are the same for any scale of production, but the equipment and the concept of individuality and special care in each stage are emphasized. The framework of the book is a block-diagram, or flow-chart of stages and operations in the course of the winemaking. The chapters are related step-by-step to the block-diagrams (white and red grapes), explaining in great detail every mode of operation in the flow-chart. I've tried to condense the most relevant data available from literature and our own experience, on winemaking processes. The book is structured so that each chapter starts with a short theoretical background, and proceeds with the practical aspects of that specific operation.

The appendix contains additional information which I've found to be very important and helpful for professional winemakers and researchers.

The old drawings illustrating the book, were taken from nineteenth century wine literature. We feel that they give the flavor and bouquet of the old tradition and spirit of winemaking.

We hope that this book will contribute to your expanding knowledge and appreciation of wine and winemaking.

"Wine-making, instead of being carried on upon scientific principles, is conducted on the old long-established principles of *rule of thumb* and *rule of mouth*; often on no principles at all; and no doubt, thousands of gallons of bad wine are made yearly, where it would have been quite as easy and a expenseless to make good wine instead of bad."

From "Chemical testing of wines and spirits",
by J.J. Griffin, F.R.C. London (1872).

"The wines of the moderns are, there is no doubt, much more perfect than those of the ancients, as far as can be discovered by any thing authentic, which has reached the present time."

From "History and description of modern wines."
by Cyrus Redding. (1833)

Dr. Yair Margalit

Dr. Yair Margalit was born in Israel, where he pursued his education in chemistry at the Israel Institute of Technology in Haifa.

His Master and Ph.D. in physical chemistry was done in the field of Nuclear Magnetic Resonance.

He joined the Israel Institute for Biological Research where he headed the physical chemistry department for five years. He was a visiting research professor at The University of California, Davis in the chemistry and enology departments, and in the physiology department in the University of Pennsylvania in Philadelphia.

His professional shift to winemaking and wine technology research began when he was asked to be (beside his academic activities) a winemaker in a new established winery in Israel. He has winemaking experience in California, Israel and Europe.

Currently he is consulting with small wineries in Israel, and teaching wine technology in the department of food engineering in the Israel Institute of Technology.

His book is the result of years of field research, practical experience, literature surveys and scientific studies.

Acknowledgements

I would like to acknowledge the persons who helped me very much during the preparation of this work:

To Professor V.L. Singleton from the Department of Enology and Viticulture at the University of California at Davis, who was my host during my stay there. From Professor Singleton I learned a lot about the chemical background, and especially enjoyed his wide and sometimes controversial views on winemaking.

To Professor R. Boulton from UCD which indirectly, but under his influence, I was exposed to his special and unique approach to the technology of winemaking.

To Zelma Long of Simi Winery in Healdsburg, California, for the discussions we had, and for giving me the opportunity to check and study all the fine details of winemaking in the winery which she is in charge of.

To Robert Pepi Jr. owner of Robert Pepi Winery in Oakville, California, and to Tony Soter, a wine consultant, for our many discussions, advice and hospitality during 1987 harvest.

To Ernie Weir, vineyard manager of Domaine Chandon winery, in Yountville, California and owner/winemaker of Hagafen Cellars in Napa for our vineyard management discussions, field trips, and valuable advice.

To Raymond Paccot, owner of La-Colombe Winery in Fechy, Switzerland, who showed me the old European tradition of small wineries.

Lastly to Jonathan Tishbi, owner of Barron Wine Cellars in Binyamina, Isreal, who trusted me and gave me the opportunity to be his first winemaker.

CONTENTS

White Grapes

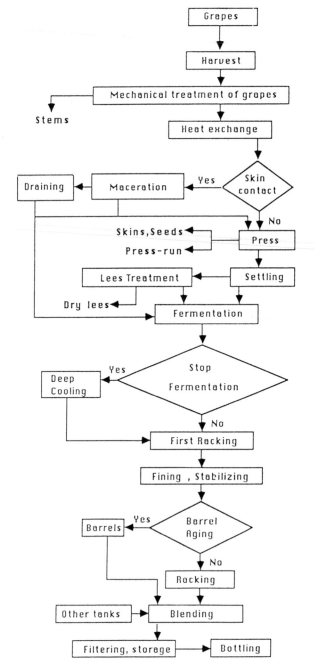

A. Grape Ripening
  1. Sugar
  2. Acids
  3. Maturity
  4. Sampling

B. Pre-harvest Oper.
C. Destemming/Crush
Q. Must Analysis
D. Sulfur-Dioxide

I. Cooling

E (1). Skin Contact

F. Free-Run/Press Juice

F (3). Lees

G. Must Corrections
H. Yeast
J. Fermentation
I. Temp. Control

K. Residual Sugar

L (1). Racking

M. Fining
Q. Wine Analysis
L (2). Stabilization
L (3). Filtration

N. Aging/Barrels

L (4). Blending
L (3). Filtration

O. Bottling
P. Quality Control

Red Grapes

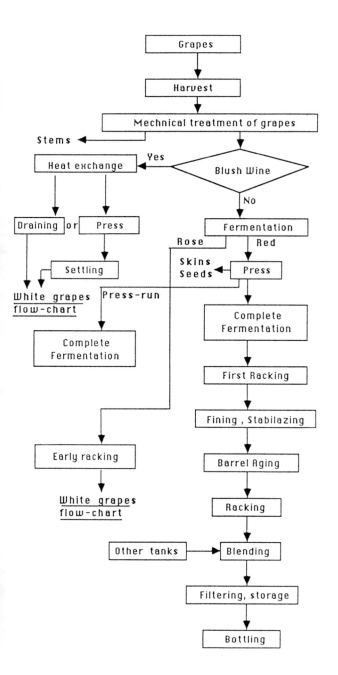

A. Grape Ripening
  1. Sugar
  2. Acids
  3. Maturity
  4. Sampling

B. Pre-harvest Oper.
C. Destemming/Crush.
R. Must Analysis
D. Sulfur-Dioxide

E (2). Skin Contact

G. Must Corrections
H. Yeast
J. Fermentation
I. Temp. Control

F. Pressing

J (4). Malo-Lactic
          Fermentation

L (1). Racking

M. Fining
R. Wine Analysis
L (2). Stabilization

N. Aging/Barrels

L (4). Blending

L (3). Filtration
R. Wine Analysis

O. Bottling
P. Quality Control

A. GRAPE RIPENING

The main object in following up the developments and changes of the grapes during the ripening period is to confirm the maturity state of the grapes, with respect to the wine we are going to produce. The maturity definition of the grapes (or more specifically the time to harvest) is not a simple and clean - cut decision; it depends on many factors, such as the variety of the grapes, the appellation (soil and climate), the weather during the last days or weeks before the prospective time of harvest, the bunch health (mold infection), the vineyard disease state, and the style of wine to be made.

The development of the berries from the time of fruit set until ripeness is roughly sketched in figure A.1. At the beginning of stage III (Veraison), the berries begin to change color and soften. Just after these changes, the growth of the berries is very fast, followed by rapid increase in the sugar concentration. Stage IV, where the grapes are over-ripe, is characterized by further softening of the berry's skin, shriveling due to water evaporation, and finally drying or mold infection. The exact time to harvest, as will be discussed in this chapter, is somewhere in the interval between the end of stage III, and up to a certain point in stage IV.

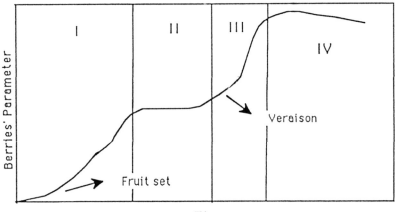

Figure A.1 - The four stages in grapes development toward maturity. The berry's parameter might be the berry weight or its sugar accumulation. Veraison is the time of color changes and beginning of skin softening.

The changes to follow-up are the sugar content, the acid content, and the pH level. Start following these changes when color begins to change. The white grapes will turn from a deep green to yellow-green, and the red grapes from green to purple-red. The change is quite rapid, a matter of a few days.

1. Sugar

The main sugars in grapes are D-glucose and L-fructose, generally in 1:1 ratio, with fluctuations of ± 30 percent, depending on the variety and maturity of the grapes. In grapes infected with *Botrytis Cinerea* mold, the ratio is in favor of fructose, which is twice as sweet as glucose. During fermentation with most yeast strains, the consumption of glucose is faster than that of fructose, and toward the end of fermentation, most of the residual sugar is fructose.

As sugar is the major soluble solid in the must (later in this chapter we address the percentage composition of the must), it's concentration measurement is based on the density of the must, assuming that the sugar is the only dry solid extract (which holds true at the range of 90-95%).

The units which are used to measure the solid concentration are:

Brix (or Balling) - percentage of soluble solids in 100 grams of solution (grams of solid/100 grams solution). This unit is in common use in the United States, Australia and many other countries.

Baume - percentage of potential alcohol, or more specifically, the potential alcohol in grams/100 ml. of wine. This unit was originally used and is still in use in France.

Oechsle - the density difference between the sample and water $(d = 1.000)$ in whole units (three digit expression):

Oechsle $= $ (density-1.000) x 1000.

For example, in a sample with $d = 1.074$, its oechsle is 74. This unit was originally used in Germany, and is still used in Germany, Switzerland, and Austria.

The Brix, which is the most used unit for sugar concentration in the wine industry worldwide, will also be used in this book. Since the

Brix (measured by density) includes all solid materials in the must (sugar, acids, salts, proteins, pigments, etc.), the approximate percentage of pure sugar content (within ± 0.5% accuracy) can be estimated. At the relevant Brix (B°) region in stages III and IV (15B° - 25B°) the pure sugar concentration may be given by:

$$\% \text{ Sugar (w/v)} = (\text{Brix} - 2.1) \times \text{density} \qquad (1)$$

Note that the % sugar is expressed by weight per volume.

The sugar is converted by the fermentation, mostly to ethyl alcohol and carbon dioxide :

$$C_6H_{12}O_6 \text{-----------} > 2\ C_2H_5OH + 2\ CO_2$$
180 gram   92 gram   88 gram

By the molar ratio of the products to the substrate (sugar), 51% of the sugar would be transformed into ethyl alcohol. In practice, part of the sugar is utilized by the yeast itself, and part is converted to other products (higher alcohols, aldehydes, esters, acids, etc.). Therefore, the molar yield of ethyl alcohol production is around 47%. In order to have a good estimation of the potential alcohol content in the finished wine (expressed in the common volume-per-volume measure), one can use the relation :

$$\text{each } 1.7\% \text{ sugar (w/v) --------} > 1\% \text{ alcohol (v/v)} \qquad (2)$$

By combining expressions (1) and (2), an expression connecting the potential alcohol to the Brix-density relation can be derived:

$$\% \text{ alcohol (v/v)} = \frac{(\text{Brix} - 2.1) \times \text{density}}{1.7} \qquad (3)$$

This expression is valid in terms of expression (1) validity, namely in the Brix range of 15B° - 25B°, within accuracy of ± 0.2% of the potential alcohol. Thus, for example a must with 22.7B° and density of 1.095 has a potential alcohol of 13.3 ± 0.2%.

The final ethanol content is also dependent on the temperature of fermentation due to evaporation losses (see chapter J.1).

Another tool for measuring the sugar concentration is the refractometer, which is based on the refractive index of sugar (linear dependence on concentration). This tool is handier than the hydrometer (which is used for the density measurements), and therefore very useful for field tests. It's values are slightly lower than those measured by the hydrometer. For details on how to measure the Brix and how to make temperature corrections see chapter R.1.

The full picture of the relationship between the must density and the various aspects of the sugar content and the potential alcohol concentration can be seen in figure A.2 .

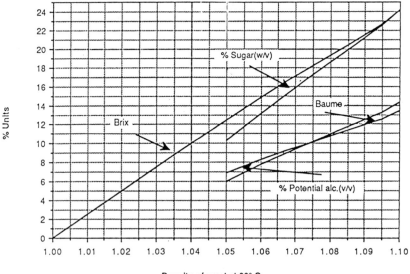

Density of must at 20° C
(assuming the density of water at 20° C is 1.000)

Figure A.2 - The relationship of Brix, % sugar, potential alcohol, and Baume, on the density of the must at 20°C. When temperature correction is needed, for all practical purposes, for each 5°C deviation from 20°C, one should add 0.001 to the density (when the must temperature is above 20°C) and subtract 0.001 from the measured density (when the must temperature is below 20°C). The correct soluble value should be read after the density correction had been made. If the Brix is read directly by the hydrometer, roughly each 5°C deviation (from 20°C), can be corrected by 0.2B° unit. (e.g. at 15°C one should subtract 0.2B° from the hydrometer reading.)

2. Acids

The main acids in must are L(+) tartaric acid and L(-) malic acid. The tartaric acid is almost unique to vines, while malic acid is quite common in many fruits.

At the beginning of ripening, the ratio of the tartaric to malic acid is almost 1:1. During the advanced maturing process, toward the end of stage III, the concentration of malic acid decreases gradually by respiration, while that of tartaric acid remains practically constant. Consequently, the reduction in acidity during maturation is due primarily to the malic acid degradation. At harvest, the tartaric/ malic ratio is generally between 2-4, depending on the variety, climate, and maturity state. The total titratable acidity in must (mainly these two acids) is in the range of 3 to 15 gram/L.

Another acid which exists in the must in a comparable amount is L-citric acid with concentration range between 0.1-0.3 gram/L, while in grapes infected by *Botrytis Cinerea*, it's concentration is usually higher, up to 0.6 gram/L.

The most important parameter reflecting the acidity in the must is the pH. The pH of grape juice at maturity runs between pH = 2.9 to 3.8. Due to the organic weak acids and the alkaline metals, mainly potassium, the must has some buffer capacity (see chapter G.1).

The acidity strength of a weak acid can be expressed by its dissociation constant :

$$Kd = \frac{[A^-][H^+]}{[AH]}$$

$[A^-]$, $[H^+]$, and $[AH]$ are the molar concentrations of the acid anion, the proton ion and the undissociated acid respectively. The acidity can also be expressed by its PKa (where $PKa = -\log Kd$).

The dissociation constants (at 25°C) and the corresponding PKa's for the most common acids in must and wine are presented in Table A.1.

Acid	Dissociation constant	PKa
tartaric	11.0×10^{-4}	2.96
malic	4.0×10^{-4}	3.40
citric	8.4×10^{-4}	3.07
lactic	1.5×10^{-4}	3.82
acetic	1.7×10^{-4}	4.76
succinic	6.5×10^{-5}	4.18
fumaric	1.0×10^{-3}	3.00
carbonic	3.5×10^{-7}	6.46
sulfurous	1.8×10^{-2}	1.77

Table A.1 - The dissociation constant and PKa at 25°C of the most common acids in must and wine. In the di-acids the values are for the first proton dissociation.

From the definition of the PKa, the following expression can be calculated:

$$PKa = -\log Kd = -\log \frac{[A^-]}{[AH]} - \log [H^+]$$

$$\boxed{PKa = pH - \log \frac{[A^-]}{[AH]}} \qquad (4)$$

From this expression and the figures in table A.1, it is clear that for example, in wine at pH 3.4, the malic acid is dissociated to its anion form at a ratio of $[A^-]/[AH] = 1$.

(from eq.4, $3.4 = 3.4 - \log [A^-]/[AH]$; ==> $\log [A^-]/[AH] = 0$; ==> $[A^-] = [AH]$). In the same wine, the tartaric acid is dissociated much more, at a ratio of $[A^-]/[AH] = 2.75$. This means that tartaric acid releases H^+ (or A^-) at this pH, almost three times more than the malic acid does. Lactic acid in this wine will dissociate less than malic acid, at a ratio of $[A^-]/[AH] = 0.4$. Other acids such as acetic, succinic, carbonic and others, are almost undissociated in this wine.

From these figures, it is clear that tartaric acid is the strongest organic acid in the must and, being at the highest concentrations is the main source of proton ions.

There is no simple correlation between the acid's concentration in the must and the pH, because of the presence of alkaline ions, mainly potassium. During the ripening period, as the malic acid concentration decreases, and the potassium ion concentration increases up to 0.5 - 1.0 gram/L, the net result of these changes is that the pH gradually increases.

As for the acidic taste, it has been found that at the same pH, the acidic taste increases in the following order: tartaric < citric < malic. This order is obvious, because it requires reduced acid concentration (e.g. less tartaric then malic) to produce a certain pH number, according to the order of the PKa's values. On the other hand, at the same acid concentration (measured by titration), the acidic taste increases in the order: citric < tartaric < malic.

The total concentration of acids in must or wine is determined by titration with sodium hydroxide solution. The common expression for the total titratable acidity (TA) value, is as tartaric acid (molecular weight $M = 150$), or as sulfuric acid ($M = 98$). For full details on acids and acidity analysis, see chapters G.1 and R.2.

3. Maturity

There are some different criteria of how to determine the maturity stage of grapes and the optimal time for harvesting. It depends very much on the grape variety, the type of wine being produced, and the climate and microclimate of the vineyard. In any method used to determine maturity, there is practically no need to start the observations before the grapes have reached 14°-16° Brix. The measurable parameters to follow up are the Brix, TA, pH and berries weight.

The ratio Brix/total acidity is a parameter that increases with time, because the sugar is gradually accumulated while the malic acid decreases. Generally, toward maturity the slope of this curve begins to be more moderate and sometimes almost parallels the time axis. The harvest should not start too late after the moderation of the curve.

Another method takes into account only the sugar development and the loss of water per berry. It is based on the product of PxS where: P = weight of a certain and constant number of berries, 200 to 500 (at any weight unit), and S = the average Brix of those berries. During ripening period, both sugar and water are accumulated in the berries. At maturation, the sugar accumulation slows down considerably and the water content begins to reduce (figure A.3). This method is quite satisfactory, although it does not take the acidity into consideration.

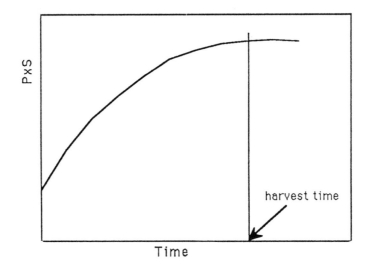

Fig. A.3 -**Brix x Weight** method for determination of grape maturity.

P = weight of constant number of berries.

S = average Brix of that sample.

Another method emphasizes the varietal character development of the grapes. In this method, the most important parameter is
the varietal aroma and flavor, and the Brix limits are a general
recommended range. The Brix limits are important only to maintain
the alcohol range in the finished wine which is typical to that variety
or style of wine. The full varietal flavor of the grape is most important
in order to achieve the maximum varietal character in that wine. This
method needs the experience and the ability to project the grapes'
flavor and taste into the finished wine. These two parameters, full
flavor and Brix range, are not necessarily met at the same time, in
which case a decision may have to be made as to which is more
important for the case at hand.

Some examples of the grapes' Brix and the wine alcohol concentration ranges in different varieties are given in table A.2 :

Variety	Brix Range	Alcohol range (%)
Sauvignon Blanc	20-23	11.0 - 12.5
Semillon	19-21	10.5 - 12.0
Chardonnay	21-23.5	12.0 - 14.0
Riesling	18-21	10.0 - 12.0
Muscat	18-20	10.0 - 11.5
Cabernet Sauvignon	21-23	12.0 - 13.5
Pinot Noir	21-22.5	12.0 - 13.0
Zinfandel	22-24	12.5 - 14.0

Table A.2 - The optimum Brix range at harvest and the corresponding alcohol concentration in the wine in some
grape varieties.

There are some more complicated methods which reflect the
ripening development, such as measuring the tartaric/malic acids
ratio, the polyphenolic compounds development, or the potassium
development, but none of these methods include a conclusive
precise assessment of the maturity timing. In cold growing regions
where grape ripening is a well known problem (e.g. Germany), the
major wine quality parameter is the Brix content of the grapes at
harvest.

Probably the widest guideline for the winemaker is to follow the very general requirements of the three basic parameters: the sugar, the total acidity, and the pH. These parameters should lie in the following ranges :

Type of wine	Brix	Total Acidity (TA)	pH
White	19 - 22	0.7 - 0.9	3.1 - 3.4
Red	21 - 23.5	0.6 - 0.8	3.3 - 3.6

If, for example, in white grapes the Brix is still low (around 18 Brix), the TA has dropped down to below 0.6, and the pH is above 3.4, one should probably wait for the Brix to rise and make the necessary corrections for the deficiency in acid and high pH. On the other hand, with the same TA and pH, if the Brix is close to 21-22, one should not wait but harvest right away.

In certain places and certain years, weather conditions (cool summer, cloudy skies) may cause a delay in the ripening, and if the delay is accompanied by rain and high humidity, the grapes will probably be infected with mold, which might get worse as the harvest is delayed. Under such conditions there is no alternative but to harvest before the full development of the sugar and flavor. The deficiency in sugar level and high acidity may be adjusted (see chapter G), but the aroma and flavor of the wine may suffer, and probably be of low-to-medium quality.

4. Sampling

The technique of sampling is very crucial in order to get a real representation of the grapes' degree of ripeness. First, the distribution of the major components in the cluster is not even. The top berries on the cluster are higher in sugar than the bottom ones. The clusters themselves are not even in their development, depending on their location on the vine, the amount of light they are exposed to, and the location of the vine in the vineyard. On top of this, the content of the berry is not uniformly distributed. Usually the sugar concentration is higher closer to the skin and drops down toward the center of the berry. The acids' concentration follows the opposite trend. All the above factors are important to realize in order to get the best representative must sample from the vineyard.

The sampling technique which might give quite a good picture on the grapes maturity is to collect 250-500 berries, each one from different cluster locations on the vine, lighted and shadowed, from both sides of the rows, and from different vines in the area. It is advisable not to collect the sample from a very large area of the vineyard (not more than 5 acres per sample), because of the variabilities due to microclimate and soil in the same vineyard. The berries should be picked as whole as possible in order not to lose their juice. For the follow-up, one should pick the same number of berries each time. The number of berries collected as a sample will reflect the range of error in that sample. Two hundred and fifty berries will set the Brix measurement within ±1 Brix range at 95% confidence, while 500 berries will set it to ±0.5 Brix range.

The berries should be weighed in order to establish the average total weight gain per berry (in water and dry components), and then crushed with a small hand-screw crusher. The must has to be separated from the skins and seeds by cheesecloth and measured for Brix, total acidity and pH. One should not hand-crush the berries in order to extract the juice, because this soft crushing leaves some parts of the berries uncrushed, and the sample does not truly represent the juice as produced using winery machinery.

The sampling and measurements should be taken once or twice a week, depending on the rate of changes occurring in the grapes' development. When the grapes are almost mature, daily sampling is recommended. For the analysis details of must see chapter R.

5. Composition and Yield

The overall weight composition of the grape clusters is :

	% (Range)	Average
stems	2 - 6	3
seeds	0 - 5	4
skins	5 - 15	12
must	75 - 85	80

The lower figure in the must range (75) is for the free-run and light press juice, and the upper one (85) includes the pressed juice.

As for the must, its composition is :

	% (Range)	Average
water	70 - 85	80
sugar	17 - 25	20
organic acids	0.3 - 1.5	0.6
other dry extract	0.3 - 1.0	0.5

The other dry extract contains proteins, amino acids, esters, alcohols, polyphenols, minerals, and aroma components.

From the wine production aspect the yields are :

White wine - about 60% of the grapes' weight (made of free-run) up to 70% (made with press-run included), ends up as white wine.

Red wine - about 65% of the grapes' weight (free-run up tp 75% (press-run included), ends up as red wine.

These figures can be expressed in a different way: from every ton of grapes (American ton), red or white, one should get about 65 to 70 cases of quality wine filled in 750 ml. bottles, at 12 bottles per case.

B. PRE-HARVEST OPERATIONS

The winery policy for the coming season should be set up before planning the major aspects of the winery operation. This should include the following points :
- Types of wines to be produced;
- Quantities of each type;
- Grapes source (quantities, quality and prices);
- Financial planning;
- Marketing prospects;
- Availability of manpower for the crush season.

1. Vineyard Management

From the veraison time up through the harvest, vineyard management has a close relation with the wine quality and wine production and should come to the winemaker's attention.

a. Chemicals used against different kinds of insects, fungus, bacteria and virus diseases.

Some chemicals may have residual activity causing health hazards, and some (especially those preventing fungus infections) may inhibit fermentation. When sulfur dust (used against powdery mildew) is carried by the grapes into the wine, it may cause production of hydrogen sulfide, which has a very unpleasant odor. The chemicals are mainly used as preventive agents during shoot and leaf growth (March - July in the northern hemisphere) and are needed much less toward the end of the summer. At this time the main concern is the bunch rot caused by different microbial agents such as *Botrytis Cinerea, Aspergillus, Penicillium, Acetobacter* and others. "Bordeaux Soup" (copper sulfate, widely used as a repellent against downy mildew in Europe), will carry copper into the must, and may later cause copper haze in the wine. (If the concentration of copper is below 0.5 mg/L, the haze will not be formed, and at the same time it may precipitate the H_2S in a "stinking" fermentation. For details on copper effects see chapters P.2 and Q.2).

In order to prevent all these chemicals from being carried onto the grapes and into the wine, their use should be terminated around the time of veraison (color changes of the grapes). The time remaining until harvest (4-8 weeks) is enough for the chemical agents to either decompose, evaporate, or be swept away by wind or rain.

b. Irrigation during the summertime when there is no rain.

If irrigation is continued up to harvest, it may dilute the dry extract of the berries and reduce the grape quality. It is believed (with no firm evidence), that the vine should struggle and be under stress during the last period of ripening in order to produce a good quality wine. In any case, extra water at this stage will be used by the vine mostly for shoot growth which does not contribute to the berrys' development. Therefore, it is generally recommended to stop irrigation at least a month before the harvest. Care should be taken in hot summers or in hot regions, because drying the vines too early may cause the leaves to turn yellow and to stop photosynthesis. Consequently, the maturity may be delayed or even worse, not come at all. Therefore, very careful inspection should be carried out with regard to the irrigation regime. Probably the best method is to irrigate small quantities once or twice a week, just enough to keep the vine from turning yellow.

c. Leaf removal and pruning.

From the time of veraison it is important that the clusters are not over-shadowed. Light exposed berries develop better flavor and higher sugar level than the shaded berries. An efficient way to create better light exposure, is to remove the leaves over the clusters level.

It is an intense labor operation, but it pays by higher grape quality. The leaf removal should start at veraison stage and be repeated when necessary.

Long shoots on the vine do not contribute to the berries development, and they create obstacles for hand harvesting as well as mechanical harvesting. It is good practice to prune long shoots, clearing the way in the vineyard and also decreasing some of the shadow areas on the vine. It is recommended that this pruning be done about 2 to 4 weeks before harvesting, depending on the vigor growth of the vines.

d. Harvest.

The traditional harvest is done by hand and for small wineries, is much preferred over mechanical harvesting. Mechanical harvesting is used mainly by large wineries because of the low availability of manpower for hand harvesting, which is seasonal hard work with low pay, and because mechanical harvesting allows picking the grapes at night when the temperatures are low, an advantage in warm regions. The main disadvantage of mechanical harvesting is that the major portion (80-90%) of the berries' skins are broken, releasing the juice into the container. Only when the crop yield is large enough is it economical to mechanically harvest. The break-even point is around four tons per acre. In lower crop yields, it is more economical to hand harvest because the machine cost per ton gets too high.

In hand harvesting, damage to the berries is minimal, and oxidation and wild fermentation is almost entirely prevented. The major factor promoting potential damage (beside the temperature) is the time allowed to pass from harvest until the processing of the grapes in the winery. It is generally assumed that a small winery receives grapes in small batches, hence from a hand harvested vineyard. It is also recommended when hand harvesting, to harvest from very early in the morning until noon to pick the grapes at cooler temperatures.

2. Preparing the winery for harvest

The summertime before the harvest is the best time to prepare the winery for the new season. The principle of preparing the winery is that everything needed during harvest should be ready, because at harvest there is no time to look for missing chemicals, fix broken machinery, leaking valves in a tank, etc.

a. Tanks.

Some of the tanks may still be full with wine from the previous year. Estimated calculations of the total volume expected to arrive and the availability of fermenting and storage tanks should be made. Extensive bottling may be required to ensure enough empty tanks for the new wine. Each tank should be checked for corrosion, valve defects and leakage problems.

b. Cooling system.

The cooling unit should be in good mechanical condition, including the cooling liquid (usually ethylene glycol) which has to be checked for proper volume in the system. All pipelines from the cooling units to the tanks and heat exchangers should be leak-free and thermally isolated. It is highly recommended to test the whole system for its total cooling capability, and every individual tank for its thermostatic control operation. These tests can be done with plain water. Detailed inspection of the cooling system might save a lot of problems later during the wine processing.

c. Barrels.

If some barrel fermentation is planned, the barrels should be readied for use by checking with water for leakage and sterilizing with sulfur dioxide. The barrel fermentation (for reasons which will be mentioned later) is best done in new barrels. The number of barrels should be carefully matched to the quantity of wine planned to ferment. A clean, cool space should be reserved for this purpose.

For details on barrel managment see chapter N.

d. Machines.

All processing machines, such as scale, hopper, conveyer, crusher/destemmer, presses, pumps, lees filter, hot water pressure machine, and fork lift should be in good mechanical working condition. The machines should be checked for any potential mechanical or electrical problems and appropriate measures taken. The hoses should be checked for leaks and enough hoses of different sizes and lengths should be ready for use.

e. Chemicals.

All chemicals used in the winemaking process should be at the winery, each one according to the amount needed with some to spare. The chemicals to be prepared are :

- Sulfur dioxide (as liquid in steel tanks, or as powdered potassium-meta-bisulfite).
- Organic acids (tartaric, malic, citric).
- Yeast (dry or culture).
- Malo-lactic culture (if needed).
- Yeast nutrient.
- Di-ammonium-phosphate.
- Fining agents (bentonite and any others needed).
- Filter pads of all pore sizes, and diatomaceous-earth (DE).
- Cleaning and sanitary materials.

f. Laboratory equipment and chemicals.

The laboratory should be ready for must and wine analysis. All the equipment for the basic analysis of Brix, total acidity, pH, color intensity, volatile acidity, sulfur dioxide, alcohol, and residual sugar should be ready including calibrated solutions. All chemicals needed for the general laboratory work should be on hand (see appendix E).

EDITOR'S NOTE: *At the time of publication, the use of di-ammonium phosphate $(NH_4)_2 HPO_4$ in wine making is controversial due to the formation of urea, a precurser of ethyl carbamate, a known carcinogen.*

g. Sanitation

A clean winery may prevent some microbiological troubles during the wine processing. Any piece of equipment which will be in contact with the must or wine (machines, tanks, hoses, barrels) should be cleaned with suitable cleansing agents (see chapter L.5) and rinsed with plenty of water. The whole floor area of the winery can be cleaned very efficiently by the hot-water pressure machine.

The sewage system of the winery must be checked for possible blockage. If there is any risk of shortage in water supply during harvest time, it is highly recommended to have a full tank of water on reserve. A winery with sudden water shortage faces a very serious problem.

C. DESTEMMING/CRUSHING

The purpose of destemming is to separate the stems from the must, because they contain very high levels of tannins, and may contribute a hard "vegy" or "green" flavor to the wine, if fermented with the must.

The purpose of crushing is to break the berries' skins allowing the release of the juice either by pressing (in white must) or by fermentation (in red). These two operations of destemming and crushing are usually done by one machine. (In some machines, the destemming is done first, and then the crushing; in others, vice-versa.)

There are different aspects in regard to these operations, for white and for red grapes.

a. White grapes

Fermentation of white must is never carried out with the skins (unlike the red). Therefore, the separation of the stems and the skins is a necessity. The destemmer/crusher followed by press is not the only method to do so. It can also be done by direct pressing the whole clusters without destemming/crushing. The idea behind pressing the whole clusters is to minimize skin contact in the must. This technique is most practical in the processing of sparkling wine and in grape varieties which have specific bitterness in the skin, such as Muscat related varieties or Semillon. (For more details see chapter E).

During the mechanical treatment of the must, the absorption of oxygen is very intensive because of the large surface area that the must is exposed to air by the mechanical operation. The oxidation (and hence browning) of white must is very fast, especially if the must temperature is not low enough (above $10°C$) and is considered detrimental to the quality of the wine.

The time taken for mechanical processing of the grapes, especially whites, should be minimal. The grapes are crushed and sulfur dioxide is added (see next chapter) as soon as possible. Recently, some wineries do not add sulfur dioxide at this stage (in white wine), in order to oxidize and polymerize some of the phenols, which may then precipitate during fermentation and fining. By doing so, the later browing of the white wine might be reduced. This early (intended) oxidation of the white must, in contrast to the attempts to prevent it, is still controversial. Later after fermentation, during wine processing, the exposure of the wine to air, from pumping leaks, filtering and the process of racking, damages the wine's quality.

If skin contact is desired, the must can be transferred through heat exchangers in order to lower the temperature to about $10°C$, and then to a drainer or to a storage tank.

If skin contact is not desired, the must is transferred through the heat exchanger directly to the press and then, to the settling tank. (On the considerations for and against skin contact, see chapter E.).

b. Red grapes

After crushing, the must is transferred to the fermentation tank to begin fermentation. The separation of the skins from the must is done later on, during or after fermentation, according to the wine-maker's judgement.

If "blush wine" (blanc de noir) is to be made from red grapes, the must is transferred from the destemmer/crusher through the heat exchanger to the press. The free-run and the press-run are generally separated and blended later, according to the desired pink color of that wine. From the press the juice is transferred to the settling tank to be treated as white wine.

The red skins of the blush juice, coming out of the press, can be added to the must of a red wine (in fermentation stage), to enhance color, tannin and body, if it is desired.

In certain cases it may be considered to use the stems to enhance the tannin level and to add complexity to the wine. This method is used on a portion of the grapes when this effect is desired, for example with Pinot Noir. The must containing the stems is blended later with the rest of the destemmed wine. The stemmed portion may vary between 20-40% according to the case.

D. SULFUR DIOXIDE

"I formerly showed that the presence of oxygen was necessary to the commencement at least of fermentation, and that it exists in the leaven or is obtained by it from the atmosphere. The effect of the sulphurous acid is to unite with this substance, for which it has a strong affinity, and thus to render the leaven powerless,   while it appears also to render it insoluble, and thus to precipitate it. The practice in common use for this end, is termed sulphuring."

From "Remarks on the Art of Making Wine" by J. Macculloch, 1829.

1. General Practice

Sulfur dioxide has been added to wine during production for many years. It was believed that sulphur dioxide had something to do with the capability of reducing oxygen, and was added after fermentation to prevent the wine from spoiling.

Nowadays, sulfur dioxide is added:

a. To protect against enzymatic oxidation of polyphenolic compounds in the must before fermentation, and prevent chemical oxidation in the wine during processing and in the bottle;

b. To protect against many kinds of microbial spoilage agents which are sensitive to its presence.

In water the SO_2 is in equilibrium with sulfurous acid anions, HSO_3^- and $SO_3^=$:

$$SO_2 + H_2O \xrightarrow{\;K_1\;} H^+ + HSO_3^- \xrightarrow{\;K_2\;} H^+ + SO_3^= \tag{5}$$

The first equilibrium constant $K_1 = \dfrac{[H^+][HSO_3^-]}{[SO_2]}$

$\log K_1 = \log [H^+] + \log [HSO_3^-]/[SO_2]$

$-\log K_1 = -\log [H^+] - \log [HSO_3^-]/[SO_2]$

(by definition: $-\log K_1 = pKa_1$; $\quad -\log [H^+] = pH$)

$pKa_1 = pH - \log [HSO_3^-]/[SO_2]$

$$===> \boxed{\log [HSO_3^-]/[SO_2] = pH - pKa_1}$$

The pKa_1 (at 25°C) of this equilibrium is 1.77, and the pKa_2 (for the second equilibrium of $SO_3^=$) is 7.20 .

By using the last expression let us check the molecular SO_2 concentration in two examples of wine pH (3.0 and 3.5) :

- At pH = 3.0

  $\log [HSO_3^-]/[SO_2] = pH - pKa_1 = 3.00 - 1.77 = 1.23$

  $[HSO_3^-]/[SO_2] = 10^{1.23}$ which is about 17 and hence about 6% $[SO_2]$.

- At pH = 3.5

  $\log [HSO_3^-]/[SO_2] = 3.5 - 1.77 = 1.73$

  $[HSO_3^-]/[SO_2] = 10^{1.73}$ which is about 54 and hence about 2% $[SO_2]$.

The concentration of $SO_3^=$ ion, for all practical purposes is negligible as can be seen from the following :

- At pH 3.0 for example, $\log [SO_3^=]/[HSO_3^-] = 3.0 - 7.2$

$===> [SO_3^=]/[HSO_3^-] = 10^{-4.2} ===>$ about 0.01% $[SO_3^=]$.

The above calculation can be summarized in a graph showing the pH dependence of the molecular SO_2 concentration (see figure D.1 line a). It becomes higher as the pH gets lower.

Although the molecular SO_2 concentration is relatively low, its antiseptic and antioxidant properties are due to this form. The $SO_3^=$ form, which is more effective as an antioxidant, had been shown above to exist in even lower concentrations (about 0.01%) in the wine pH.

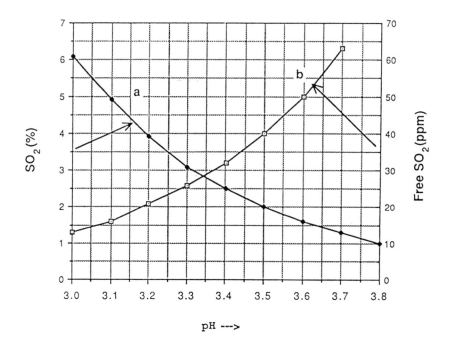

Figure D.1 - The relative ratio (%) of molecular SO_2 from the total
sulfur-dioxide in solution vs. the pH (in water at 25°C).
Line **a** is the relative concentration of molecular SO_2.
Line **b** is the free SO_2 needed in solution (ppm), in order
to maintain 0.8 ppm of molecular SO_2 at given pH.

The effective SO_2 concentration range as antiseptic agent is at
about 0.8 ppm of molecular SO_2. In order to use the data in figure
D.1, let's take an example. Assume that the wine pH is 3.3, and after
racking, the measured free sulfur-dioxide was 15 ppm and the total
SO_2 is 35 ppm. According to figure D.1 at pH = 3.3, 26 ppm of free
sulfur-dioxide is needed to maintain 0.8 ppm of molecular SO_2 in
solution (line b). We have to add 26 - 15 = 9 ppm more free SO_2. It is
roughly estimated that in healthy and not rotten grapes, at total SO_2
concentration below 50 - 60 ppm, any SO_2 added to the wine will be
divided equally to free and bound SO_2. So, to have approximately 0.8
ppm of molecular SO_2 in this wine, one should have to add 18 - 20
ppm of sulfurdioxide. At higher total SO_2 concentrations (above 60
ppm), any new addition of SO_2 should be considered roughly as free
SO_2 addition.

These figures are general guidance, and the exact concentration of the free SO_2 should be analytically measured after the addition (see chapter R.4).

The sulfurous anion (HSO_3^-) reacts irreversibly with aldehydes and proteins in the wine and becomes bound or fixed. This part of the sulfur-dioxide is no longer active. It may also react reversibly with some phenolic compounds in the must, which make this part of the sulfur-dioxide a reservoir of free SO_2 which may be released when its concentration gets too low.

Besides protection, in white wines the SO_2 preserves and magnifies the fruitiness of the wine.

A very small amount of sulfur-dioxide is formed during fermentation by the yeast, up to about 10 ppm. This sulfur-dioxide is usually fixed by acetaldehyde as soon as it is formed. So, if one should analyze SO_2 presence in the wine (total), he would find some, even if it was not added.

The concentrations of SO_2 which are used in the wine industry are usually much lower than those which may be believed to cause some health hazards. The maximum legal level of total SO_2 concentration in many countries and in the United States is 350 ppm (350 mg/L). It is worthwhile mentioning that some people are allergic to SO_2 at even very low concentrations. A new law in the U.S. (1987) requires printed notification on the label that the wine contains sulfites for any wine in which the total SO_2 is greater than 10 ppm. The average threshold of tasting free SO_2 is about 50 ppm.

For healthy grapes the quantity of SO_2 recommended to be added before fermentation, as soon as possible after crushing, is 30 - 60 ppm for white wines and 0 - 80 ppm for red wines.

For moldy and rotten grapes, more SO_2 should be added (50% - 100% more) because these grapes contain higher concentration of aldehydes which will fix a higher amount of the added SO_2. The total SO_2 measured after fermentation would in general be lower then the SO_2 added, because part of it was absorbed by the settling yeast. After the first racking, the SO_2 level should be analyzed (total and free) and more should be added to the wine to maintain about 0.8 ppm of molecular SO_2. During bottling, it is recommended to have

20 - 30 ppm of free sulfur-dioxide in the wine. Any addition should be calculated according to figure D.1 .

These figures of SO_2 addition to must are noticeably lower than those of a few years ago in the U.S. but are still much the same in many wineries in Europe.

For white must, some wineries do not add any SO_2 during crushing (especially Chardonnay), and only after the first racking do they add the minimal amount 30 - 50 ppm (see chapter C).

In red wines, no addition of SO_2 during crushing is recommended when ML-fermentation is aimed to follow the regular fermentation. Only after the ML-fermentation has taken place, the SO_2 is added during the first racking. This principle holds in any case of intended ML-fermentation (e.g. in Chardonnay or Sauvignon Blanc when a certain portion of the must is going through ML-fermentation).

In red wine, after the ML-fermentation is over, the recommended addition of SO_2 during the first racking is 50 - 60 ppm. After the second racking, a second addition can be made according to the analysis of the free and total SO_2 and by following the above example.

It is important to mention that the 20 - 30 ppm of free SO_2 which is enough to give protection in white wines (pH in the range of 3.1 - 3.4), is *not* enough in most cases of red wines, where the pH range is generally higher (3.4-3.6). At pH 3.5 for example, the concentration of free SO_2 which is needed to maintain 0.8 ppm of molecular SO_2 is 40 ppm, and at pH 3.6 it is oven higher (50 ppm). To maintain such high free SO_2 in the wine, the total SO_2 would be considerably higher. To avoid that high SO_2 level, some wineries do not maintain enough free SO_2 as required by the pH of the wine, knowing that the wine is at risk of microbial spoilage (mainly *Brettanomyces*); instead they keep 20 - 30 ppm free SO_2 level and, in order to minimize the risk, check frequently for possible spoilage. Probably over a period of years, better bottle aging of red wine is achieved with low addition of SO_2.

In late harvest Botrytised grapes, it is necessary to add higher quantities of SO_2 because of the very high concentration of aldehydes in the grapes. The addition is made right after pressing, in concentrations of about 80 - 120 ppm.

In California's commercial wines, a study has recently shown an average of 17 mg/L free and 125 mg/L of total SO_2 in white wines. The minimum and maximum of these values were found to range between 5 mg/L of free SO_2 up to 240 mg/L of total SO_2.

To enhance the antioxidant capacity of SO_2 in white must, one can very successfully use ascorbic acid (vitamin C). It can be added to the must to protect against browning right after crushing at amounts in the range of 5 - 20 gram/HL.

2. Methods of addition

The sulfur dioxide can be administered to the must or wine by one of three convenient methods :

a. By adding potassium-metabisulfite ($K_2S_2O_5$) which decomposes at the acidic media of the must to SO_2, at about 57% of its weight. It is convenient to prepare a solution of the salt (e.g. 87.5 gram per one liter of water) and use this solution according to the ratio :

$$\boxed{\text{Volume of } SO_2 \text{ solution (ml)} = 2 \times HL \times ppm} \qquad (6)$$

where : HL - is the wine volume in Hectoliters.

ppm - is the SO_2 concentration in ppm.

e.g. if 40 ppm addition is needed for 3,000 gallons (which is about 12,000 liters = 120 HL) of must, then 2x120x40 = 9,800 ml = 9.8L of the solution should be used.

The stock solution of the potassium-metabisulfite should be kept in a closed vessel and be used up within about a week following its preparation (or be recalibrated for the SO_2 content).

It is also possible to weigh the potassium metabisulfite for each addition, dissolve it in a small volume of water and add it directly to the must or wine.

In this case :

> 17.5 gram potassium-metabisulfite/HL $= 100$ ppm   (7)

b. By adding solution of pure SO_2 into the wine. The SO_2 is commercially available in liquid form, and is kept in steel tanks. The solubility of SO_2 in water depends inversely on the temperature (at constant pressure). Figure D.2 shows its maximum solubility values vs. the temperature. In order to prepare SO_2 solution, one should bubble the gas from the tank into cold water (15 - 20°C). The concentration of the SO_2 in solution can be determined by measuring the density with a hydrometer. Figure D.3 gives the SO_2 concentration (in gram/100 ml) vs. the density at 20°C. As for temperature corrections, when the actual density measurement is made at a temperature other than 20°C (68°F), one should subtract (when the temperature is below 20°C) or add (when the temperature is above 20°C), the corrected value to the actual density reading, and find the percentage of SO_2 from the graph. (E.g. if the density measured at 25°C was 1.026, then 0.002 should be added to it, to get the corrected density at 20°C (1.028), and the SO_2 percentage reads from the graph to be 6%).

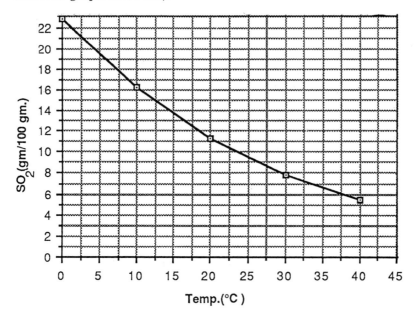

Figure D.2 - Solubility of SO_2 (at saturation) in water at atmospheric pressure, in grams of SO_2 per 100 grams of water.

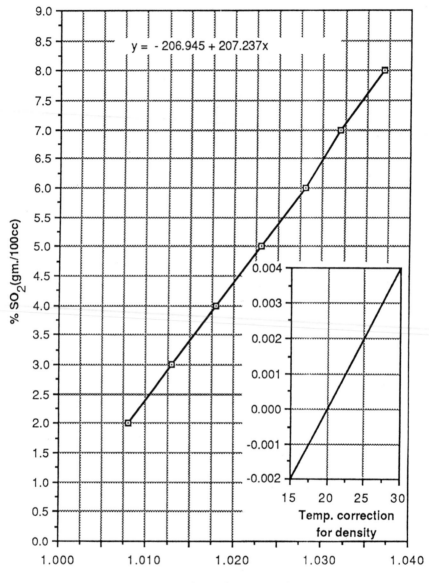

Density of solution at 20 °C.

Figure D.3 - Density of SO_2 solution in water at 20°C (grams SO_2 in 100 ml. of solution). For temperature correction refer to inner Figure.

A convenient formula to use for adding pure SO_2 solution is:

$$SO_2 \text{ solution (liter)} = \frac{(\text{ppm } SO_2)}{100 \times (SO_2\%)} \times HL \qquad (8)$$

where : $(\text{ppm } SO_2)$ - is the ppm of SO_2 needed for adding.

$(SO_2\%)$ - is the concentration of the SO_2 solution.

HL - Hectoliters of wine.

For example, assume our SO_2 solution is 4.8%, and we want to add 60 ppm SO_2 to 20 HL of wine. Then: $\frac{60}{100 \times 4.8} \times 20 = 2.5L$ of the SO_2 solution should be added.

This method of adding SO_2 is very convenient and does not add potassium to the wine as it does in the previous method. As the SO_2 tends to evaporate from the stock solution, its concentration must be checked from time to time, by density measurement or by SO_2 analysis (see chapter R.4).

c. The third method is to add liquid SO_2 from the steel tank directly to the must by SO_2 doser. There are many kinds of dosers on the market, each with its specialized instructions for use.

The traditional and old method of introducing SO_2 into the wine was by burning sulfur in a closed vessel and pumping the gas into the wine. In barrel maintenance (chapter N), when the barrel is clean and not in use, it is advisable to burn sulfur sticks (about 10 grams) in the barrel and close the bung of the barrel. This will protect the barrel against mold and all kinds of wine spoilage micro-organisms. This procedure should be repeated every two to three months. The sulfur has to be burned over a special container so that if and when it melts, it will not leave elemental sulfur in the barrel, which can sometimes be transformed in the wine into hydrogen-sulfide. Instead of burning sulfur, it is also possible to release SO_2 gas from a steel tank into the barrel for two-three seconds and close the bung.

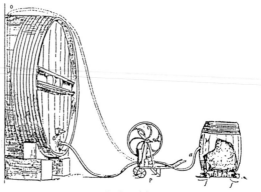

Soufrage à la pompe.

E. SKIN CONTACT (MACERATION)

Skin contact is one of the major functions in the winemaking process. It influences the type, character, aging period, and general quality of the wine. Some of the varietal flavor and most of the color and tannin compounds in grapes are stored in the skin. Their total amounts vary in ranges of 2 - 5 gram/kg in red grapes, and 0.2 - 0.5 gram/kg in white grapes. All these compounds are defined as phenolic compounds, derived from phenyl alcohol. For details on the chemical structure of the phenolic compounds, see Appendix B. Roughly, they may be divided into three functional groups, responsible for color (pigments), astringency (tannins), and flavor.

The pigments contain mostly flavanols anthocyanins (cyanidin, peonidin, malvidin, petunidin, and delphinidin), which give the wine its red color. The color is pH dependent and changes from deep-red at low pH to green-blue at about pH 5 - 6, with no sharp color transformation. They are generally bound to sugar molecules as glucosides.

One remark must be made here with significant viticulture importance. It has been found that the anthocyanin glucosides are bound to one sugar molecule (monoglucoside) in the *V. Vinifera* varieties, where in the native American varieties *V. Labrusca*, they are bound to two sugar molecules (di-glucoside). This di-glucoside character is transformed genetically in all varieties which are crossed hybrids with *V. Labrusca*. A quite simple chromatographic test for mono and di-glucoside malvidin, which is one of the most common anthocyanin in grapes, can verify whether the grapes are pure *V. Vinifera* or *V. Labrusca* cross-hybrid. (It must be added that this conclusion is not cut and dry, because there are some isolated exceptions).

Tannins are the polymeric forms of the phenolic derivatives of anthocyanins and also of benzoic acid derivatives (catechin, leuco-cyanidin, gallic acid, vanillic acid and others).

Most polymers are structured from two up to six or seven monomers, containing many -OH sites which are responsible for the "dry" feeling or astringency of red wine.

The third group, the flavor compounds are mainly derived from cinnamic acid derivatives (cafeic acid, coumaric acid, ferulic acid), and partly from the flavanols group.

The functional grouping is not strictly diverse, and some compounds may be colored and tannined, or tannined and flavored at the same time. The intensity and diversion of these compounds in the skin depend upon many factors such as variety, maturity state, climate, soil and cultivar management. This chapter deals with the extraction of these compounds from the skins into the must.

In the vinification process, the extraction of the phenolic compounds depends on the variety of the grapes, as well as the temperature, alcohol concentration, sulfur dioxide concentration, and time of contact. The extraction is proportionately related to all four of these factors. It is faster at higher temperatures, at higher alcohol and sulfur dioxide concentrations, and its concentration increases with length of contact (maceration). The total amount found in the finished wine is about 1 - 2 gram/L in young red wine, and 0.1 - 0.4 gram/L in white. In dealing with skin contact, we will separate the discussion on white and red grapes.

1. White Grapes

In white grapes, the skin contact can be categorized into three groups; no skin contact, short contact, and long contact.

In no-contact treatment, the must is pressed immediately after destemming and crushing. The most extreme practice of this minimal skin contact is achieved by pressing the whole clusters without crushing. This technique is used mostly in sparkling wine production, where neutral basic wine with minimal character is desired.

The short skin contact is done between 1 to 4 hours, where long skin contact can last for up to 24 hours. In these cases, the must is left with the skins after crushing, in order to extract more varietal aroma, knowing that there is also a gain of some undesirable bitterness from the flavonoid compounds which later might oxidize and turn yellow-brown.

When skin contact is desired, the must should be sulfited right after crushing, mixed well, and cooled to 10 to 15°C. The major effect of temperature on skin contact is the increasing rate of the total phenolic compounds extracted with increasing temperature. On the other hand lowering the temperature at this stage is necessary in order to reduce the oxidation potential of the must.

The time of skin contact depends on the variety, the ripeness of the grapes, and the kind of wine to be made. Some winemakers prefer minimal skin contact to get fresher, non-astringent white wine, while others prefer more varietal character, richer in aroma, with certain tannin levels.

For aging white wine, it is better to have some skin contact where, through aging, the phenolic compounds will contribute to the bouquet of the wine. When the grapes are fully ripened, the time for skin contact can be shorter. However, if they are unripened enough and skin contact is longer, care should be taken because the must and the wine made of it may become "green" or get a "leafy" flavor which might be undesired.

After skin contact followed by pressing, the must can either be inoculated to start fermentation, or be treated for lees separation. The last operation can be done either by gravitational settling for 12 to 24 hours or by centrifugation. In the gravitational settling the lees settle down in the tank to form a dense layer whose thickness depends on the pressing technique used (see next chapter). The clear must is racked off from the lees layer to start the fermentation. The lees left over can be filtered by the lees filter and added to the rest of the must (chapter F.3). Centrifugation, on the other hand, can reduce the solid content of the must to about 1% solids at a very fast rate. The musts' flow through the centrifuge is continuous, with interruptions from time to time, to discharge the solids. The centrifuge can also be used at any other stage of processing, mainly after fermentation to clear off the wine from the yeast lees. When centrifuging wine, care should be taken to prevent air from coming into the centrifuge (by using a nitrogen environment), because this process introduces a lot of gas into the liquid.

After lees separation, it is advisable to add 10 - 20 gram/HL of di-ammonium-phosphate as nitrogen nutrient to prevent fermentation problems (especially if completely dry wine is desired).

A somewhat different technique to treat the white must is to use a drainer (juice separator). The drainer is a vertical tank with an inner central or side screen, where the clear juice can flow through, leaving the skins behind (see figure E.1). The drainer is filled with must directly from the crusher or through the heat exchanger, and the must is left in for a certain period, according to the desired level of skin contact. The juice is separated from the skins through the screen, either by gravity pressure of the must, or by CO_2 pressure (up to 7 p.s.i.) exerted from a CO_2 tank. The principle of the drainer is that the pomace itself functions as a "filtering" media, reducing the solid particle content of the free-run juice. After the free-run juice has run out to a tank, the rest of the must is discharged into a press to extract the rest of the juice. The pressed juice is combined with the drained juice in the settling tank, where the total solids content is lower compared with a must that has not been drained.

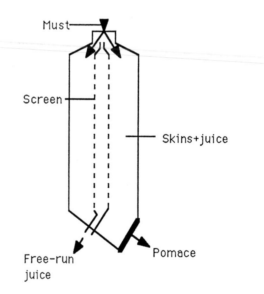

Fig. E.1 - Drainer (juice separator)

The whole process of draining (loading, waiting, draining the free-run, and discharging the must to the press), may take from about one hour, when minimal skin contact is desired, up to many hours as needed when long skin contact is desired.

Usually two or more drainers can serve one press. The drainers function also as a storage buffer between the destemmer/crusher and

the press. The drained juice contains about 0.5-2% of solids, compared to 5-10% in the usual press separation which needs a settling
time to reduce this high content of solids.

By using the draining technique for white must, one can point
out two operational advantages. First, by lowering the must load
coming out from the drainer into the press by about 25-40% of its
original volume, saving press time, where each cycle takes about
three to four hours.

The second advantage is the possibility of crushing any load of
grapes coming into the winery right away, without waiting for the
press to finish its cycle.

A remark is made here concerning the Muscat varieties (such as
Muscat of Alexandria, Muscat Canelli, Sylvaner, Symphony,
Emerald Riesling) which may develop a typical bitterness in their
wines. Minimal skin contact may reduce this problem. Pressing the
whole clusters without crushing may be the best method for such
cases. On the other hand, in Chardonnay, skin contact is generally
practiced, where in other varieties it depends on the case and on the
wines' style.

2. Red Grapes

The extraction rate of the color and tannin from the skins can be
seen in Figure E.2.

The color pigments which are mainly monomeric anthocyanins
are extracted faster than the polymeric tannin molecules. Around
zero Brix there is no increase in color extraction, therefore in regard
to color there is no need for any more skin contact. Separation of the
skins from the must can be done at any time along the extraction line.
If the separation is done shortly after the beginning of fermentation,
the result would be a good red color with minimal amount of tannins,
and the wine may be consumed very young. If separation of the skins
is made after a long period of contact, the tannin extraction will be
high and longer time will be required for the wine to age.

In red grapes, there are four kinds of skin contact treatment
which will reflect the type of wine being made:

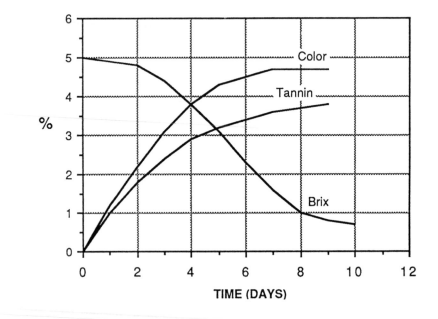

Fig. E.2 -   General picture of the extraction rate of color and
             tannins from the skin during fermentation.

a. No skin contact - immediate separation of the skins from the
   juice.

This wine, which may be almost white or slightly pink in color,
will be called "blanc de noir" or "blush" wine. This wine is considered
white wine, and its' processing is carried out accordingly.

The "blush" wine, which has become very popular in recent
years in the United States, is a light, fresh, fruity wine, and usually
off-dry with about 1-3% residual sugar. The main problem with blush
wine is that the bright pink color ages in a short time (months) to
brown-pink, which does not appeal well to the customer. The
varieties which have little of that problem are Cabernet Sauvignon
and Zinfandel. A possible solution to minimize this problem in other
varieties is to avoid the use of SO_2 at crushing before fermentation,
allowing part of the pigments to oxidize and polymerize, which settle
down later during and after fermentation. At bottling, the color of
that wine will be more stable.

b. Short skin contact without fermentation.

The contact may last a couple of hours up to 24 hours. This "blush" wine may also be called light "Rosé," and its processing is as white wine.

c. Short skin contact during fermentation.

This wine starts to ferment with the skins as red wine, but in about 24 hours, the skins are separated, and the wine continues to ferment as white wine at a cool temperature. This wine will also be called "Rosé" wine, with a more intense pink color and more tannins than the previous one.

d. Long skin contact during fermentation.

This is the red wine production, where the duration of the skin contact may last from a minimum of 3 to 4 days for light red wine, and up to 14 to 21 days for a heavy, tannic, longer-aged wine. The variety, the specific grapes, the style of the wine, and the aging time are the factors which determine the duration of the skin contact during and after fermentation.

The most light red wine production is done by a technique called "carbonic maceration", in which the grapes are fermented without being crushed and destemmed (whole clusters). The free-run of the *fermented must* is separated through a screen in the fermentation tank, and the result is a good colored red wine, with minimal astringent tannins. This wine can be consumed very young as has been done for many years in the Beaujolais region in France, where the wine is released in November (2 to 3 months after harvest).

Some varieties are intense in color and tannins (Carignane, Cabernet Sauvignon, Zinfandel, Petite Sirah), where others (especially Pinot Noir) have light pigments and low tannin concentration. The intensity also varies from season to season, and the decision when to separate the skins has to be made in accordance with each wine.

One of the methods to enhance low color intensity is thermo-vinification, namely pre-heating the grapes or must, prior to fermentation. The heat destroys the skins' cells membranes, releasing the pigments, tannins and other substances into the must. The yield of extraction by the heat is temperature dependent in the range of 40-80°C.

Thermovinification can be done either by heating the must after crushing, or by heating the grapes (whole clusters) before crushing (by hot air or steam). The second method has the advantage of heating mainly the skins with minimal heating of the pulp inside the berries. For wines that are to be consumed young, and when the color intensity is a problem, the thermovinification may help to enhance it. For long-aging wines, the extra color extracted by heating will polymerize and precipitate after a long time, and there will be no color advantage after several years of bottle aging.

A better method to enhance the color intensity in wine may be to blend it with a small portion of a variety (Zinfandel, Alicante, Petite Sirah) with a high color intensity.

In Pinot Noir, where the tannin level is low, some wineries ferment some portions of the must with the stems, and blend this portion with the rest (see chapter C). The blend is higher in tannins and somewhat more complex.

To summarize this subject, the very general principles should be remembered: the pigments are extracted more quickly than the tannins, the must should be checked during fermentation and differentiation should be made between the color intensity and the tannin level in accordance with the kind of wine desired.

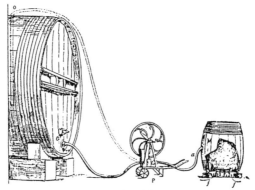

Soufrage à la pompe.

F. FREE-RUN AND PRESS JUICE

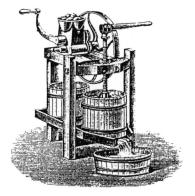

1. General View

Pressing of grapes is done either on the unfermented must for white wine and "blush wine" production or during fermentation for red wine and rosé production. The juice (or wine) which comes out of the press by using minimal pressure is called free-run, where the heavier pressed juice (or wine) is called press-run. The unfermented free-run must is about 60 - 70% of the total extractable juice, where the fermented free-run wine is about 70 - 75% of the total wine volume. This is because during fermentation the pulp cells die and the cells' membranes are not active anymore, making for an easier flow of the cells' content.

The constituent composition of the free and pressed run is different. In the press-run (of must or wine), there are more phenolic and polyphenolic compounds (pigments and tannins), less total acidity, higher potassium concentration and therefore higher pH. In pressed wine, there is also higher volatile acidity. All these parameters are related to the pressure level exerted during pressing and the type of press used.

In general and based on all parameters, better quality wine is made from free-run, where the use of press-run has to be considered according to its particular quality. It is advisable to separate the press-run and process it in a different tank. If it has good quality, it can be blended later with the free-run wine in such a proportion to get the highest quality possible. The press-run also allows the wine-maker one more possible dimension to enrich his wine in color (in red wine), tannins, and complexity.

Grape juice (as many other fruit juices) contains pectins. These polymeric compounds tend to create coloid coagulation in the wine which may contribute later on an instability factor. Press-juice contains more pectin that does free run juice. Hydrolyzation of the pectins during juice processing will prevent this potential problem. Wine or grape juice treated by pectolytic enzymes will clarify faster and will be easier fined. It is recommended for white or red wines as well. The depolymerization of the pectins is done by pectolytic enzymes which are commercially available.

The enzymes are tolerable to must conditions. Its pH optimum activity is between 3.5 to 5.0. Its optimum temperature activity is at $50°C$, but it is active between $5-60°C$. Sulfur-dioxide is an inhibitor to its activity, but it can tolerate SO_2 concentration up to 500 ppm. The enzyme also tolerable to all wines alcohol levels, so it can be used also on finished wine.

The range of enzyme addition is 2.5 - 5.0 gr/HL of must. When added to the must right after destemming/crushing it should be well mixed.

2. Presses

Presses can be divided into two major categories; batch presses and continuous presses. In the first category the oldest and simplest press is the vertical basket press (see figure F.1) which is usually made of wood, where the direct pomace pressure (up to 7 kg/cm²) is exerted either manually by a screw-jack or mechanically (hydraulic or pneumatic), compressing the pomace downwards.

By pressing the pomace the juice runs out through the basket rim and is collected from its base. The basic problem with this kind of press is that the pressure is exerted on the outer section of the pomace, while the inner and especially the center section remains partially unpressed with higher content of juice. Also, unloading the pomace after pressing, namely breaking the compressed "cake", is difficult and requires hard work.

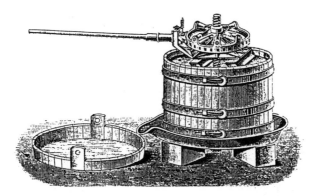

Fig. F.1 - Old vertical basket press.

A modification of this kind of press is a horizontal press, made of stainless steel where the pressure is exerted by two plates moving from the edges of the cylinder, pushing the pomace toward the center and compressing it (see figure F.2). The pressure is released from time to time, while the cylinder rotates slowly to break the pomace "cake", enabling more homogeneous pressure.

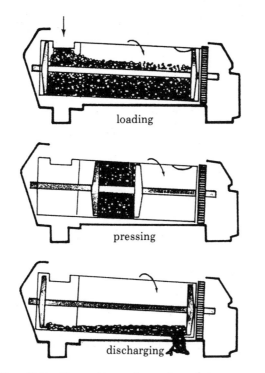

loading

pressing

discharging

Fig. F.2 - Operation of moving plates press.

A much more sophisticated and modern version of this press exerts the pressure by inflating an air sac (bladder) mounted to the side of the cylinder (see figure F.3). In the older type of inflating

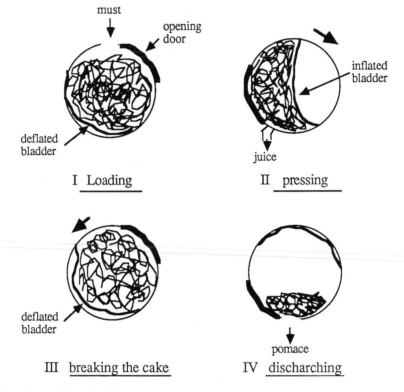

Fig. F.3 - The bladder press operation.
Steps I and IV are done once at each operation.
Steps II and III are repeated in many cycles at increasing air pressure in the baldder.

press, the air sac is centered in the drum, pushing the pomace towards the walls. The pressure in the bladder press is built up gradually in cycles where each cycle is composed of pressing the must while rotating the drum, then releasing the pressure while changing direction of rotation to the opposite direction (to break the pomace "cake"), and pressing again. The pressing time and the reverse rotation time in the cycle are set manually prior to operation. Normal cycling time is 2-5 minutes. The press operates in three cycling blocks according to the air pressure in the bladder (see figure F.4). The number of cycles in each block section is determined by the

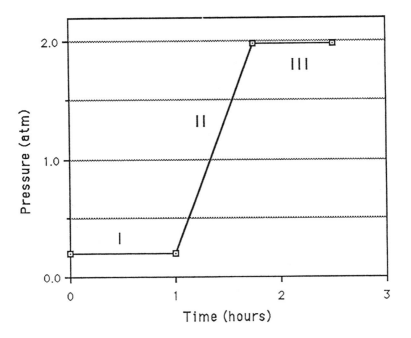

Fig. F.4 - Typical pressure profile in the bladder press.

press operator and is automatically controlled and programmed. In section I at constant low pressure (about 0.2 bar) the number of cycles is about 10 to 15. In section II the pressure is incremently increased at each cycle by 0.2 bar, up to the maximum pressure, which is pre-set by the operator (about 2 bars). In section III where the pressure in each cycle is the maximum pressure set, their number is about 5 to 10 cycles.

The total pressing time (includes loading, pressing and discharging) takes about 2 - 4 hours.

Because of the relatively low pressure exerted on the pomace and the gentle mechanical treatment of the skins, the solid content in the juice is relatively low.

The distinction between the free-run and press juice is determined at a certain pressure level in section II, or at the beginning of section III. In some wineries, with white must the decision when to separate the free-run from the press juice is made by tasting the juice coming out of the press. The parameters for judging are the high tannin level and low acidity (flat taste).

In a case where the skins are thin and slippery, the screens inside the drum are usually plugged, causing difficulties in the press operation (this phenomena is typical, for example, in Semillon). Good practice in such a case is to add a portion of the stems (from the destemmer/crusher machine) to the must when loading the press (layer of must, layer of stems and so on). The existence of the stems prevents plugging of the screens.

All types of presses described above are batch type. They must be loaded and discharged at each batch. The batch quantity may run from about 2 HL (small basket press) up to 7 - 8 tons in the bigger horizontal presses.

The second category is the continuous presses which are based on an "infinite Archimedes screw" that press the pomace very slowly to an outlet whose opening size is controlled. This enables the operator to set the exerted pressure on the pomace as desired (see figure F.5). The advantage of this type of press is that it is continuous, saving the loading and discharging operation. The main and serious disadvantage of the screw-type press is that it causes tearing and breaking of the skins' tissue, leading to more solids in the juice. All skin components are higher in must coming out of such a press.

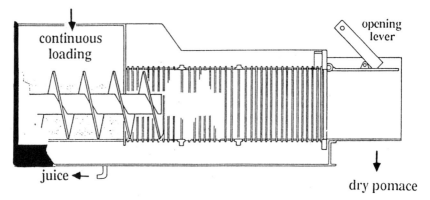

Fig. F.5 - Continuous press operation.

A continuous press is not recommended for pressing white must, if recommended at all.

3. Lees

The lees are the solid particles suspended in the must after crushing and/or pressing. This term is used also to describe the sediment that settles before the first racking or after fining, where it contains mainly yeast cells or fining particles. When leaving the white must to settle for 12 to 24 hours, depending on the particles' size and quantity, the settling lees create a dense layer (sludge) of variable thickness, from 5-10% of the liquid volume.

The juice above the lees is racked off throuth the racking valve in the tank, (see Figure L.1), and the only way to save the juice trapped in the lees layer is to extract it by mechanical force. This can be done either by lees filter, or by centrifuge.

The lees filter is very effective in clearing the juice from it's lees. The filter contains 20-40 plates which support thick cloth pads. The plates with the pads are packed together tightly in a row, at very high mechanical pressure (hydraulic pressure of about 300-500 bars on 40x40 cm plates). At the entrance of the filter there is a special high pressure pump which forces the juice to flow through the pads.

To facilitate the filtration of the dense sludge of lees, the juice is mixed well with diatomaceus-earth (DE) before entering the filter. The DE is a finely ground mineral powder which increases the surface area of the filtering pads and enables the lees to combine during filtering in a multi-layer texture with the DE, rather than one layer on the pad surface without the DE.

The mixing ratio of DE to juice is 2-4 Kg/HL, depending on the lees density. More DE is needed when the lees density is higher. The first batch of mixing can be 3-4 Kg/HL, and when this batch is filtered, and a new batch of juice is mixed with DE, the DE/juice ratio is cut to half of the first one, and if a third batch is prepared, the ratio is a quarter of the first addition.

The DE dust is not healthy to breathe and it is highly recommended that the operator wear a dust mask while mixing the DE with the must.

At the beginning of filtering, the flow is easy and no pressure is developed. The first batch of the mixed juice and DE is used to build the DE "cake" between the pads, protecting them from being plugged by the fine lees particles. As the volume of liquid flowing through increases, more DE + lees particles are deposited, increasing the flow resistance. Pressure gradually builds up by the filter pump and the increased resistance. The maximum pressure in most filters is 10 bars. The pump, which is pressure controlled, stops at the maximum pressure for a couple of seconds, till the pressure drops down, and starts again. When the pumping intervals between off and on exceed 10-15 seconds, the filter is probably plugged and the filtration can be stopped.

The filtering capacity of a lees filter with 40x40 cm plates is about 30-50 liters/plate, so a typical lees filter with 20 plates can filter 600-1,000 liters of must before it will be plugged.

Some useful hints for the lees filtering operation :

• At breakdown of the filter plates to clean it, the shape of the "cake" reflects the quality of the filtering. A "good cake" (dry, dense and easy to remove) shows the proper ratio between DE/must and good pad condition. When the cake is soft, not homogeneous and sticky, this means that the filtering was not efficiently done, the DE/must ratio was too low, or the cloth pads were almost plugged by lees particles.

• Good washing of the pads with water after each use is necessary to prevent microorganism growth. Also, after a couple of filterings, the pads get partially clotted and it is necessary to take them off from the plates to machine wash them.

• Before closing and compressing the plates for new filtering, wet the cloth pads with water.

• Check the edges of the pads before closing the plates, to assure that no bent pad is caught between the plates. If this happens, the filter will leak when the filtering pressure rises.

• The first 20-40 liters of filtered juice has an earthy taste (from the DE). Transfer it back to the pre-filtered juice.

• From time to time open the high pressure pump valve to release the DE that accumulates there.

When dealing with the lees left over after racking (yeast lees, fining lees), in general, the thickness of that lees layer is small and it is simply drained off.

G. MUST CORRECTIONS

1. Acidity

When previously discussing the acidity aspects of must in wine, it was noted that the two parameters involved, namely total acidity and pH, are not simply correlated.

Differentiation should also be made between total acidity and titratable acidity. The total acidity is the sum of all organic acids in the wine (or must), including their salts. The titratable acidity is the available acidity of the wine as free H^+ ions. The total acidity is higher then the titratable acidity, because part of the organic acids are neutralized by the various cations, mainly potassium. It is clear why the enzymatic measurement of the total acidity will show higher results than the titratable ones. In this book (and in other texts too), the titratable acidity is traditionally called total acidity (TA).

A few words about the buffer capacity of must. The buffer capacity of must, which is the response of the pH changes to addition of strong base (or acid), depends on the total titratable acidity (of the week organic acids) and on the main cation concentration in the must (potassium). The higher the total acidity and the potassium concentration, the higher the buffer capacity (with no simple relation).

A titration curve of a typical must, and of a strong acid for comparison is shown in figure G.1 .

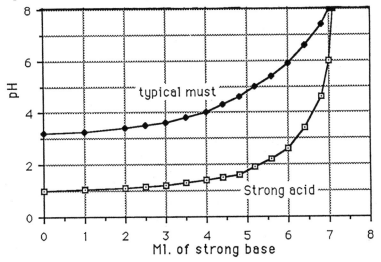

Fig. G.1 - Titration curve of equal equivlents of a typical must and strong acid by strong base.

The buffer capacity of must is in the range of $(4-6) \times 10^{-2}$ equivalent/L per pH unit, meaning that in order to change the pH of 1L of must in one pH unit, it is necessary to add $(4-6) \times 10^{-2}$ equivalents of strong base (in pH range up to about pH 5).

One should adjust both the acidity and/or the pH, according to the must at hand and the desired type of wine.

The sour taste is more dependent on the concentration of acids (total acidity) than on the proton ion concentration (pH). As a general guide, the preferred values of the must would lie in the range of 0.7 - 0.9% total acidity and pH = 3.1 - 3.4 for white wine and 0.6 - 0.8% and pH = 3.3 - 3.6 for red wine. In sweet or semi-sweet wines, it is better to have the acidity in the higher range, and the pH at the lower one, in order to balance the sweetness.

It should also be kept in mind that the total acidity may be reduced during the wine processing by malolactic fermentation and by cold stabilization, (see chapters J.4 and L.2).

The acidity problem can be divided into two categories: deficiency and surplus of acid.

The more severe and common case in white must is the lack of acid rather than the surplus of it, especially in warm regions. The advantages of correcting the acidity, when it is in deficiency, are: more controlled and steady fermentation, a reduction of the yeast autolysis at the end of fermentation and inhibition of bacterial spoilage. It also preserves the flavor and the aroma of white wine, and is necessary for the balance and general quality of the wine. The acid correction should be done before starting the fermentation, because of the above reasons regarding fermentation and because any major correction later on will greatly affect the wines' stability.

The techniques used for acid correction when it is deficient are: blending with high acid must, addition of acids, and by ion exchanging.

i. Blending with other must (high acidity and low pH) is the best natural and elegant method, when it is possible. However, in many cases, it is not practical. It may be quite difficult to find the right grapes with the desired pH and total acidity, at exactly the time of crushing and starting of fermentation. There are other factors which should be considered in blending, which are not necessarily dependent only on the pH and total acidicy, especially when blending with another variety.

ii. The most practical and frequently used method is by acid adjustment. The options are to use tartaric, malic, citric and fumaric acid; each has it's advantages and disadvantages.

a. Tartaric acid is the strongest of the three (lowest PKa), and it would need less than the others to get the same result in lowering the pH. On the other hand, a large part of the acid added will precipitate later during cold stabilization. However, by cold precipitation of the potassium-bitartarate, the potassium concentration is reduced, which leads to lower pH value. If high pH is the major problem of the must, the addition of tartaric acid and precipitating the surplus of it by cold stabilization will improve the pH value of the wine considerably. (On the solubility of potassium-bitartarate vs. the temperature, see figure L.2).

b. The main reason for acid deficiency in must is the reduction of malic acid during ripening period by respiration. The addition of malic acid into the must can be looked at as a way of restoring the original balance between tartaric and malic acids in the grapes. Malic acid, which is less acidic than tartaric acid, will not precipitate in cold temperature, but it may be transformed into lactic acid by malolactic fermentation. When the commercial DL-malic acid is used, only the L-malic acid isomer is consumed during the malolactic fermentation, leaving the D-malic isomer untouched, adding its contribution to the total acidity.

c. Citric acid can be added to expand the acid taste and to prevent possible iron haze (by complexing the iron). However, during ML-fermentation, it may be partially converted into acetic acid. So, addition of citric acid can be done safely only when ML-fermentation is inhibited because it isn't desired in that particular wine, and measures have been taken to prevent it (chapter J.4), or it can be added after the ML-fermentation is completed.

d. Fumaric acid is another option which can be used for acidification. Its PKa is practically the same as tartaric acid (table A.1). This acid is also very efficient in inhibition of ML-fermentation at concentrations of about 300 - 500 mg/L. The main problem with fumaric acid is its low solubility in water or water/alcohol solutions. The solubility in water is 5 grams/L, 23 grams/L and 98 grams/L in

20°C, 60°C and 100°C respectively. The crystals of the acid are difficult to dissolve. The best way to use this acid is to prepare hot solution of about 50 - 80 grams/L and to add it hot into the must or wine. The exact volume of the solution must be calculated according to its concentration, the must volume and the desired concentration of the acid in the wine. The legal limit of fumaric acid in the U.S. is 3 grams/L. Fumaric acid has some harsh taste and its addition should be done with care.

The amount of acid needed to correct the acidity deficiency depends on the total acidity, the pH, and the buffer capacity of the must. In most cases when using tartaric acid at a pH range of 3.5 ± 0.2, every 1 gram/liter added (0.1% TA), will lower the pH by about 0.1 ± 0.02 pH units. The (-) sign is at pH < 3.5 and the (+) sign at pH > 3.5. For malic or citric acid, each 1 gram/liter added will lower the pH about 0.08 ± 0.02 pH units, at the same pH range as above. This is a general guideline. One should try, in the laboratory, to determine the right quantity per liter in order to get the optimal TA and pH according to the must at hand.

It is probably better to use tartaric acid and small proportions of malic acid before fermentation and, if needed, to add citric acid after the first or second racking. It should be remembered that the maximum presence of citric acid in wine is limited in many countries to about 0.5 gram/liter. Its' natural presence is quite low (about 0.05 gram/liter), so it is advisable not to add more than 0.4 gram/liter.

In a case where the TA is high enough, but the pH is too high (high potassium content), one may consider using phosphoric acid in order to lower the pH without adding too much to the total acidity. The use of phosphoric acid is common practice in the food industry, although it is not used in the wine industry. Its use is not recommended (also illegal in the U.S. as a wine additive). It gives the must and the wine a "watery" or low body feeling.

iii. The ion exchange method exchanges the potassium cation with hydrogen and is a good technical solution to the above case. The ensuing exchange is very efficient and almost all the potassium is replaced by hydrogen ions, lowering the pH substantially. So, only a portion of the wine is transferred through the ion exchanger column, and then blended with the rest of the wine. The result of such blending will lower the total pH of the wine.

Deacidification of the must is the other side of the problem, where the must acidity is too high (above 10 gram/L) and the pH is too low (below 3.0). This problem is usually the result of under ripe grapes which had to be harvested because of some viticultural difficulties, such as weather conditions, vine diseases or bunch rot. The problem is quite common in Germany, parts of France, Switzerland and other cool regions.

The taste of the wine from such must is sour, tart and un-balanced.

The low pH will also inhibit ML-fermentation when it is desired (in red wine).

In order to correct the excess acidity, some measures can be taken, namely blending with low acidity must, chemical treatment, ML-fermentation and cold stabilization.

i. The best method is to blend with low acidity and high pH must. Again, the same consideration mentioned earlier in this chapter on low acidity correction by blending is applicable here as well.

ii. The other way to reduce the acidity of the must (or wine) is by using chemicals. In principle, the neutralization of the excess acid by base is not good, because the salts formed will give the wine a salty taste and sometimes an off-flavor. It is preferable to remove the excess acid by precipitation with calcium carbonate, potassium carbonate, or potassium tartarate. All will react with the tartaric acid to produce hard soluble tartarate salts which will precipitate, and lower the TA:

$$CaCO_3 + 2H^+ + 2HT^- \; \text{---------->} \; Ca(HT)_2 + CO_2 + H_2O$$
$$K_2CO_3 + 2H^+ + 2HT^- \; \text{---------->} \; 2KHT + CO_2 + H_2O \qquad (9)$$
$$K_2T + H^+ + HT^- \; \text{---------->} \; 2KHT$$

HT^- is the first dissociated tartaric anion.

T stands for tartarate anion.

The reaction can also include the second dissociated anion ($T^=$) to give CaT. Also insoluble double salts of tartaric and malic acids can be formed.

The quantities of carbonate or tartarate which have to be added must be determined in the laboratory before addition. As general guidance, experience shows that the following quantities of deacidification agents, which upon addition to the must will increase the pH value by 0.1 pH unit, are:

$CaCO_3$ - (0.3 - 0.4) gram/L
K_2CO_3 - (0.4 - 0.5) gram/L
K_2T - 1.5 gram/L

It is important to remember that although the salts formed are insoluble and precipitate, the Ca^{++} or K^+ ions concentration in the deacidified must or wine is increased sometimes to the point where instability problems may arise (chapter P.2).

iii. ML-fermentation is a good and "natural" treatment for reducing acidity in wine. In this fermentation, the malic di-acid is transformed into the lactic mono-acid:

$$COOH\text{-}CH(OH)\text{-}CH_2\text{-}COOH \text{ --------}>$$
$$CH_3\text{-}CH(OH)\text{-}COOH + CO_2 + H_2O \tag{10}$$

reducing the acidity concentration of malic acid by half. Also the PKa of lactic acid is higher than the PKa of malic acid (table A.1) which means that it is less acidity. The net result of the ML-fermentation is a reduction in the acidity and an increase in the pH value by several tenths of pH unit.

A major problem in using this method is when the pH is below 3.1 - 3.2, where it is difficult to carry on the fermentation under these pHs. In such a case where the pH is lower than this value, it can be brought up to pH = 3.2 by other methods (such as carbonate addition), and then inoculated with ML culture. For details see chapter J.4 .

iv. Cold stabilization is an important operation during the process of winemaking. In this operation the excess of tartaric acid is precipitated as potassium-bitartarate salt, lowering the pH (because of the potassium precipitation). So, cold stabilization reduces the total acidity on one hand, and reduces the pH on the other. This controversial aspect of the cold stabilization on deacidification has to be considered carefully in each actual case.

It is not recommended to treat chemically the excess acidity of the wine before cold stabilization takes place. Only after this operation does one know to what extent the excess acid is a real problem to the wine.

For details on cold stabilization, see chapter L.2 .

2. Sugar

There are several reasons for the deficiency of sugar concentration in the grapes at harvest time, and all of them lead to unripened grapes. When this happens, leaving the grapes on the vine for further maturation and delaying the harvest may not improve the sugar accumulation in certain cases, or will cause damage to the grapes in others. As a result of unripened grapes there is not enough sugar to ferment into the normal alcohol concentration range in wine (10 - 13% V/V). Usually these grapes also contain excess acidity and in many cases, suffer from more severe problem which cannot be corrected with additives; namely, not enough varietal flavor and aroma. There is nothing wrong from the quality aspect with making sugar or acid corrections; however, because these grapes are unripened, the quality may suffer from lack of varietal distinction.

In cold regions (Germany, North of France, Switzerland) the addition of sugar is necessary almost every season, and it is legal to do so. The regulations usually require a certain minimum of natural Brix (around 15-16 B°), and notification to the authorities about the sugaring (chaptalization), and how much. In some other regions (e.g. California) it is absolutely forbidden.

The sugar addition can be done with cane sugar (sucrose) or with grape concentrate (about 70% sugar). When sucrose is added, it has to be hydrolysed to mono-sugars before it can be fermented by the yeast. It is naturally done by inversion enzyme contained in the must.

The quantity that has to be added depends on the sugar level in the must and on the desired alcohol concentration. This can be deducted from equation 2 (chapter A.1). From this relation the quantity of sugar addition can be expressed in any convenient unit:

> 0.92 Kg sugar/HL -----------> 1 Brix unit

> 1.65 Kg sugar/HL -----------> 1 Baume unit   (11)

> 2.1 Kg sugar/HL -----------> 10 Oechsle units

For example 0.92 Kg of sugar added to 1HL (100L) of must will increase the Brix by 1 unit.

If cane sugar is used, it is best to dissolve it in boiling water to make very highly concentrated syrup (about 80 - 90%), before adding to the must. The volume needed can be calculated from eq.11 and the syrup concentration.

To prepare a concentrate syrup, boil 25 liters of water, and add 50 Kg of sugar while mixing. After a few minutes the syrup will clear up. The syrup volume will be about 55 Liters at about 90 Brix.

To measure its exact concentration, dilute a sample by 5 (1 volume of syrup + 4 volume of water), and measure its Brix with a 16B° - 24B° range hydrometer. The syrup concentration would be the Brix reading multiplied by the dilution factor (5).

Knowing that the Brix unit is grams of solid dissolved in 100 grams of solution (chapter A), the volume to be added can be calculated by the following example. Assume that the syrup is 83B° and you want to add 3 more Brix to the must. Then add :

$$\frac{92}{83} \times 3 = 3.3 \text{ liter of syrup/HL of must}$$

It is recommended to add the syrup while it is still warm (40 - 50°C), before the sugar recrystalizes.

If grape concentrate is used, the sugar concentration is known by the producer, and the same calculation can be made.

The addition of sugar has to be done at the beginning of fermentation, when the yeast is vital and strongly active. If added at the end of fermentation, it may sometimes get stuck and difficult to achieve dryness.

H. YEAST

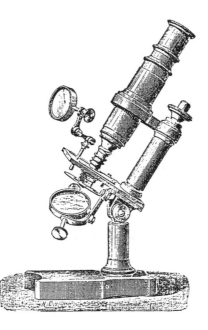

1. Wine Yeast

Surprising as it may seem, people have made wine for thousands of years without knowing why and how the grape's sweetness is transformed into alcohol. Only in Pasteur's time, the mid-nineteenth century, was it found that it is a microbiological process, conducted by yeast. If the must is left by itself, the "natural" yeast found on the grapes and in the winery will start on their own fermentation. Over many years, through research and trial and error, the more favorable yeasts for wine production were selected. The known number of yeast species related to winemaking is about 150, categorized into about twenty different genus. The most important for wine production are those belonging to the *Saccharomyces* genus *cerevisiae* and *bayanus* species.

The *Saccharomyces cerevisiae* and *Sacchanomyces bayanus* species (the first name designates the genus and the second, the species) are subclassified into different strains of yeast whose differences are based on our exploitation of their special characteristics in the winemaking process, like:

    a. Tolerance to different conditions, such as temperature, alcohol and sulfur dioxide.

    b. Different by-products during fermentation.

    c. Flocculation capability (colloidal or granular) which affects their sedimentation after fermentation is over, and hence the wine's clarification.

The most abundant "natural yeast" belongs to another genus, *Kloeckera apiculata*, which remains active only up to 4 - 5% of alcohol. At the opposite extreme of tolerance to alcohol, is the *Saccharomyces bayanus* species which tolerates and is active up to 17 - 18% alcohol.

The commercial yeast strains of *Saccharomyces cerevisiae* (also called *Saccharomyces ellipsideus*) and *Saccharomyces bayanus* are sold in dry vacuum packages or as liquid cultures. The shelf life of the dry yeast is over a year if kept unopened in cool (not cold) temperatures, whereas the liquid culture yeast's shelf life, under the same conditions, is about a month. As said before, if no yeast culture is added to the must, it will start fermenting by itself on its own natural yeasts as is still practiced in many French Chateaux. However, the use of one dominant yeast strain is most beneficial as it is more consistent, faster and tailored to a winemaker's exact needs.

Some of the commercially available strains and their specific characteristics are :

Montrachet (U.C. Davis #522) - *Saccharomyces cerevisiae strain* all purpose yeast for red and white wines. At moderate temperatures (15-30 °C) fermentation is fast, aggressive, foamy and easy. Sometimes tends to produce hydrogen sulfide, especially when sulfur dust is present. Good tolerance to sulfur-dioxide and medium tolerance to alcohol concentration. Low production of volatile acidity and good complex flavor.

Pasteur Champagne (U.C. Davis #595) - Despite the name, it is not used for sparkling wines. An all-purpose yeast, especially for white wine. Moderately fast, and being *Saccharomyces bayanus* it has good alcohol tolerance, usually completing the fermentation to dryness. A strong living yeast, which can be used to regenerate stuck fermentation. Good tolerance to SO_2. Recommended for fruit wines. Good flavor.

Pasteur Red - Fast and easy fermentation. Recommended for red wines with full body, such as Cabernet Sauvignon. Sometimes it requires cooling of the fermented must in order to keep the temperature from getting too high.

Pasteur White - Recommended for white, not fruity wines. Tends to foam, unless fermented at a cool temperature.

Epernay 2 - *Saccharomyces Cerevisiae strain.* Recommended for white fruity wines. Fermentation is relatively slow, especially at cool temperatures. Sensitive to very low temperatures, suitable for wines intended to have some residual sugar by halting the fermentation at the desired sugar level. Sensitive to lack of nutrition.

California Champagne - (U.C. Davis #505) *Saccharomyces Bayanus strain.* Its main character is that it settles easily, suitable especially for sparkling wines (champagne method). Leaves a yeasty flavor which is typical for these kinds of wines.

Prise de Mousse - Being *Sacchanomyces Bayanus*, it has high tolerence to alcohol concentration and to sulfur dioxide. Because of the last two reasons, it is very suitable for late harvest high sugar grapes which have been treated with a high dose of SO_2 (grapes infected by *Botrytis Cinerea*). Good fermentation at a low temperature which is most suitable for white wines. Low production of hydrogen-sulfide and extremely low foam production. Good flocculation properties with good activity under pressure, which makes this strain suitable for champagne method production. Very good starter for stuck fermentation. Widely used in recent years in the United States.

Tokay - *Saccharomyces Cerevisiae strain,* with very high alcohol tolerance, up to 17% (V/V). Sensitive to cold temperature, but high tolerance to SO_2.

2. Starting Fermentation

The yeast community multiplies itself by budding at a rate which is controlled by the media conditions, such as: sugar concentration, temperature, alcohol concentration, nutrient, oxygen and chemicals present. The population at fully active fermentation process is about 100 million cells per milliliter. The population changes during the process, can be followed by using a microscope with a magnifying power of about 500 - 1000, allowing good vision on the yeast cells which are about 1 - 2 micron (10^{-3} mm) in diameter.

In order to start a good and healthy fermentation, one has to add enough living yeast cells to the must. In using dry yeast, which is very convenient, the quantity needed is about 10 - 20 gram/HL, equal to about half a million to one million cells per milliliter of must. The yeast should be hydrated first by water/must mixture (10% must) at temperatures of 38-40 °C, for about 15 to 20 minutes. The ratio of the dry yeast to liquid is about 1:10 (1Kg dry yeast to 10 liters of water/must mixture). Be aware that in 10-20 minutes, the newly hydrated yeast may overflow from its container as a result of very fast fermenting activity. After hydration, it should be introduced directly into the must. Mixing is not necessary.

When using liquid culture, one should prepare a starter by heating grape juice diluted by half with water, up to boiling in order to sterilize it. When cooled down to room temperature, add 80 - 100 ppm SO_2 and the yeast culture, and mix vigorously in order to aerate it; cover the vessel and mix it once or twice later on. The starter will be ready in about a day or two (at room temperature). The volume ratio of the starter to the must should be about 2% in order to get good starting fermentation.

Regarding the temperature at the starting stage, there are no problems with red wines, as the temperature is about 22 - 30 °C. For white wines, where the fermentation temperature is about 10 - 14 °C, it may take quite a long time before the fermentation starts. It is recommended, therefore, to bring the must to temperatures of about 15 - 17 °C, then add the yeast or starter, and wait for good signs of fermentation (a reduction of one - two Brix), when the yeast population has grown enough to carry on the cold fermentation. Then reduce the temperature to the desired range (10 - 14 °C). If the yeast in use is highly tolerable to cold fermentation (e.g. Prise de Mousse), the inoculation can be done at the fermentation temperature (10 - 12 °C).

A final word regarding safety during fermentation. **Never add additional must to a fermenting tank from the bottom valves !!** The result may look like an explosion. The reason is that the fermenting must is supersaturated with CO_2 gas (especially during cold fermentation), and if some perturbation happens to the liquid (like the waves of incoming liquid from the bottom), it may release the extra dissolved CO_2 at once, which looks like an explosion.

If addition of must is necessary to a fermenting tank, it can be added slowly and carefully to the **top** of the tank.

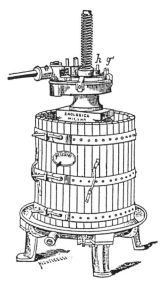

I. COOLING AND TEMPERATURE CONTROL

"The temperature is one of the external circumstances which has the greatest share in influencing the act of fermentation. It has been considered, that a heat of about the 54° of Fahrenheit's scale is that which is most favourable to this process. There is nevertheless some latitude to be allowed; but in a temperature either very cold or very hot it does not take place at all."

From "Remarks on the Art of Making Wine" by J. Macculloch 1829.

1. Cooling

Controlling the temperature during fermentation is one of the most important factors in good winemaking. The range of temperatures in which the yeast is active and ferments is between $10°C$ - $35°C$, $(50°F$ - $95°F)$. At the high temperature range, the fermentation starts faster, but as the alcohol concentration increases, it slows down. At $35°C$, sometimes it may even stop, leaving some residual sugar unfermented. At low to moderate temperatures, the fermentation starts slowly, proceeds more moderately, and generally it will go on to dryness. The time lapse between inoculation and the first signs of fermentation can take about one, two, four and seven days for temperatures of $35°C$, $25°C$, $15°C$ and $10°C$, respectively. The lag time depends also on other factors which will be discussed in chapters J.1 and J.2.

The alcohol is an inhibitor for the yeast growth, and its inhibition effect is greater at high temperatures. It also evaporates more through the CO_2 bubbles, at higher temperatures, which makes the alcohol yield higher at low temperature fermentation. The variation in alcohol content may be as much as 1% absolute alcohol difference if the fermentation is carried on at temperatures of 20°C and 10°C. So it is expected that the wine will be fuller bodied if fermented at lower temperatures. Also, at lower temperatures, the fruitiness of the grapes is better preserved, by reducing evaporation of volatile aroma components from the must. Also the volatile acidity level at low temperature fermentation has been found to be lower than at high temperatures.

For all the above reasons and for practical considerations, it has been accepted that the preferred fermentation temperature for white wines is between 8°C - 14°C (46°F - 57°F). This is also true for rosé wines, and white wines made from red grapes ("blush wine"), as all of them are considered white wines. For red wines, the fermentation temperature should be higher, between 22°C - 28°C (72°F - 82°F), for two reasons: better color and tannin extraction, and because less fruitiness is desired in red wines.

2. Temperature Control

Fermentation is a heat source process. Although the energy yield of this aerobic process is poor, and the yeast itself uses up part of this energy, there is still some energy released as heat. Experimental calorimetry measurements have shown that one hectoliter of must, at about 22 Brix will release 3,000 Kcal as heat. This means that if the container is isolated without releasing the heat, the temperature could rise (theoretically) by 30°C.

Practically, the natural heat losses through the container's walls and the CO_2 evolution through the liquid reduce the expected theoretical temperature. This reduction depends on the total volume of the must (or more accurate, on the *walls surface/volume ratio*), on the container material (cement, wood or stainless steel), and on the environmental temperature (temperature difference between the fermenting must and the surroundings).

In order to carry on the fermentation at the desired temperature, one should control it by means of a mechanical cooling system. The modern technique uses cooling liquid (usually ethylene glycol) flowing in a cooling jacket around the tank. The liquid is cooled by a cooling system which has to have the cooling capacity needed for all the winery functions. The tank's jacket is generally built as one or two strips at 2/3 and 1/3 of the stainless steel tank's height. The heat flows from the liquid to the tank's walls. It is very difficult to build a cooling device in cement or wooden tanks. The temperature can be controlled individually at each stainless steel tank by a temperature sensor placed in the tank. The sensor should be long enough so it will read the liquid temperature far away from the walls where it is coldest.

Because there are two temperature gradients in the tank- one radial, from the center toward the walls (the walls are cooled by the jacket liquid), and the second from top to the bottom of the tank (due to temperature density differences)- the temperature control does not necessarily control the whole volume at the same temperature. During fermentation these gradients are reduced by the fermentation turbulence, but later they are more steady. One should be aware of this fact in regard to the real temperatures in the tank.

In some places in Europe the fermentation cooling is done simply by dripping cold water on the outside of the stainless tank walls, where the flow rate of the water serves as the temperature control.

Another method is to use an air conditioned room which is much less efficient, because of the low heat capacity of air. In barrel fermentation (of white wine) this is the only method to somewhat control the fermentation temperature.

In red wines the fermentation may take between one and two weeks, whereas with whites, it may take three to six weeks to finish. With white wine, if toward the end the fermentation seems to be slowing down very much or even stuck, it is sometimes advisable to raise the temperature to 15°C - 18°C in order to complete the fermentation.

After fermentation and during the rest of the winemaking process (racking, fining, barrel, etc.) up to bottling, the temperature should be controlled at 20 ± 5°C for reds, preferably at the lower range, and 12 - 15°C for white wines.

When quick cooling is needed, such as when the must has to stay on the skins for several hours at a cool temperature, there are cooling machines which allow reduction of the must temperature by 5 - 15°C at a very high flow rate through a heat exchanger. In any fast cooling with a heat exchanger, it is highly recommended to pass the must (or wine) through the cooling system to another tank, rather than to pump it back to the same tank. By returning the cooled liquid back to the same tank, the temperature gradient between the cooling system and the liquid gradually gets smaller and smaller, with lower and lower cooling yield. When the liquid is transferred to another tank, the gradient remains constant, making the cooling much more quick and efficient.

J. FERMENTATION

On the Nature and Causes of Fermentation, and on the Substances engaged in it.

"Since it is from the chemical changes which take place in this mysterious process, and from the new combinations which are formed amongst the elements of the different bodies which have been described, that wine is obtained from this mass of heterogeneous substances; it is essential to consider the nature of fermentation so far as any light has hitherto been thrown upon it."

From "Remarks on the Art of Making Wine" by J. Macculloch, 1829

1. Alcohol Fermentation

The main purpose of fermentation is to convert sugar to alcohol. This process, which is carried out by the wine yeast, is the major process of making wine and its understanding and control are basic for good winemaking. From the biochemical point of view, the process had been studied in great detail.

The reader who is interested in the details of alcoholic fermentation and the possibilities of side reactions of the process is referred to biochemistry texts.

Here we shall deal only with the general aspects and external conditions which influence and control fermentation without going into biochemical details.

From the chemical point of view, alcoholic fermentation can be described as a three-step process:

a. Breaking down the six carbon sugars - glucose and fructose through phosphorylation into a three-carbon molecule, phosphoglyceraldehyde.
b. Decarboxylation of the three-carbon molecule into two-carbon acetaldehyd and carbon-dioxide. This is the carbon dioxide source in fermentation.
c. Reduction of the acetaldehyde to ethyl alcohol, as an end product.

The overall conversion is:

One sugar molecule ====> 2 ethyl alcohol + 2 CO_2 (12)
(molecular weight 180)   (M.W. 46)   (M.W. 44)

About half of the energy released in this process is used by the yeast, and the rest is dissipated as heat. The energy yield of the process is very inefficient from the yeast's point of view.

Without side reaction, the theoretical yield of alcohol production would be 51% of the sugar weight (92 grams of alcohol out of 180 grams of sugar). In practice, because of side reactions, evaporation and unfermented sugar, it may end up with 90 - 95% of the theoretical value, which is about 45 - 48% of the sugar weight.

The total volume of carbon-dioxide gas released during the fermentation process is about fifty times the must volume (at atmospheric pressure, room temperature and at 18 - 20 Brix).

After inoculating the must, the fermentation propagation can be viewed in terms of yeast cell growth or by sugar consumption (see figure J.1).

The yeast growth pattern can be divided into four stages:

A - Lag period, where the yeast acclimates to the must (sugar concentration, pH, temperature, SO_2).

B - Exponential growth, where about 50% of the sugar is consumed in a relatively short time.

C - Stationary phase, where nutrients become scarce, and inhibition by alcohol, fatty acids and higher alcohols begin to slow down the growth.

D - Decline phase, where severe shortages of nutrient and sugar and poisoning by the fermentation products occur.

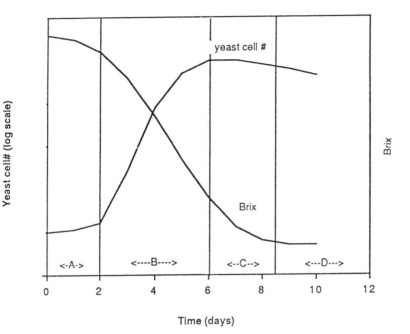

Fig. J.1 - Yeast growth pattern.

The Brix changes follow in accordance with yeast growth. Two of the factors which influence the fermentation rate, yeast nutrient and temperature, will be discussed in the following sections of this chapter. In addition, we shall mention two other factors, sugar and alcohol, as both of them inhibit fermentation at certain concentrations. At high concentrations of sugar, starting the fermentation might be difficult, and in botrytised grapes with 30 - 40 B°, the fermentation usually stops at 7 - 10% alcohol, because the combination of alcohol and high concentration of sugar is toxic for the yeast growth.

The alcohol inhibition power depends on the yeast strain. The "natural yeast", *Kloechera apiculata*, are rendered inactive at relatively low concentrations (4% - 5%) of alcohol. Most *Saccharomyces cervisiae* strains tolerate 13 - 15% of alcohol concentration (depending on many other factors), and the *Saccharomyces bayanus* can ferment up to 17 - 18% alcohol.

There are some differences in the equipment for fermenting white wines and fermenting reds, because of the skin (cap) management needed in the reds, and temperature controlled tanks needed for the whites.

The volume of the white must should not be more than 90% of the tank volume, as to not overflow the foam when the fermentation is in its maximum rate.

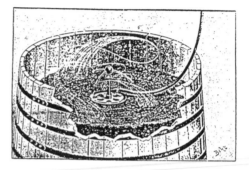

The red must containing the skins should be placed in a fermentation tank equipped with a pumping-over device which can spray the must taken from the bottom of the tank over the floating cap of skins. This pumping-over is necessary for color and tannin extraction, and should be done three to four times a day at the beginning of fermentation (up to about a quarter of the original Brix), and then twice a day until pressing or until the cap settles down (in the case when pressing is delayed for long skin contact).

Cap management can also be done mechanically, by punching down the cap in an open tank, but this is an elaborate and difficult procedure. The must volume of reds should not be more than 75 - 80% of the tank volume, in order not to overflow when the cap is floating and pushed up by the carbon-dioxide gas.

The recomended method to start the fermentation is to use dry yeast strain, selected for the specific needs of the must at hand. For details see chapter H.1. For instructions how to prepare the starter from the dry yeast see chapter H.2 .

The fermentation process has to be followed carefully by the winemaker from the beginning to end. During fermentation, the major indicator of the fermentation rate is the Brix change. For details on how to measure it and how to make corrections, see chapters A.1 and R.1.

Any stage of the fermentation can also be checked microscopically, by counting the yeast cells (alive and dead) and by watching the normal budding of the yeast's cells (see chapter R.10). It is also necessary to follow temperature changes during fermentation and prevent any drastic changes due to failure of the cooling system.

If something goes wrong, action should be taken as soon as possible to prevent serious problems which can develop later on and affect the wine quality. One of the most common problems during fermentation is "stinking fermentation". Experienced winemakers can detect it even at very early stages, by smelling the fermenting tanks daily. The stinkiness is caused by hydrogen-sulfide (H_2S) produced by the yeast (see this chapter, section 3). This phenomena is quite common and is mainly caused by nitrogen deficiency. When discovered at an early stage of fermentation, the best treatment is to add 100-200 ppm of diammonium-phosphate (DAP) to the fermenting tank. The stinkiness may dissipate in one day if discovered and treated early. If the stinkiness is discovered at the end of the fermentation, it is of no use to add DAP, and the best treatment is to aerate the wine by racking. Aerial racking can be done simply by opening the bottom valve of the tank and pumping the wine from that container into another open tank. If the H_2S has not yet been transformed into mercaptan (by interaction with the alcohol), the aeration will be very helpful. At a later stage of the wine processing the hydrogen sulfide problem can be treated only by fining with copper (chapter Q.2).

Some wineries with old traditions of winemaking put a piece of copper bar in the must container at the output of the press (in white wine). This brief contact with the copper bar dissolves some copper ions in the must, which then precipitate as copper sulfide when the H_2S is formed, eliminating the stinky odor. This method is quite efficient but has the risk of introducing too much copper into the wine, which may cause copper haze instability later on. The easiest way to prevent or minimize this problem is first, avoid any sulfur dust on the grapes by good vineyard management, and second by adding DAP to the must before fermentation (at the time of inoculation). The recommended quantity is 150-200 ppm.

Toward the end of fermentation, analysis of alcohol, residual sugar, volatile acidity, total acidity, pH and sulfur-dioxide, gives the winemaker the basic information needed on the processed wine.

2. Stuck Fermentation

Sometimes the fermentation gets stuck or even does not start at all. The reasons may be:

a. Lack of Oxygen - The yeast needs oxygen in order to bud, although the whole process of fermentation itself is anaerobic. This is important at the early stage of fermentation. In order to aerate the must, one can pump about a quarter to a third of the must over the top of the upper surface of the tank. A good practice in white must fermentation is to rack the fermenting must two days after inoculation into another tank. This racking will aerate the must right at the end of the lag period and also homogenize it (yeast culture and nutrient).

b. Lack of Nutrient - Especially nitrogen; some winemakers regularly add DAP (diammonium-phosphate $(NH_4)_2HPO_4$) to the must before starting fermentation. It is generally not needed in red wine fermentation, but in white must when the juice is clear after lees racking, it may be necessary to add 10 - 20 grams/HL of DAP to prevent fermentation problems such as stuck fermentation or "stinky fermentation." If the fermentation is stuck, it might be useful to add 10 - 15 grams/HL of DAP and mix it by aerial racking. Sometimes it may be helpful to add the amino acids L-phenylalanine (2 grams/HL) and thiamine (50 mg/HL). There are commercial yeast nutrients, sometimes called yeast-extract, which contain mixtures of all the necessary ingredients at the right proportions. Only in rare cases of stuck fermentation is it advisable to use such mixtures. When the yeast extract is used the quantities are about 20-40 gram/HL.

c. Unviable yeast - If fermentation does not start after the above measures, a new yeast starter should be introduced. To start a stuck fermentation, the new starter must be acclimated to the alcohol and the environment of the stuck must. In such cases the starter should be handled as follows: After preparing the starter as described in chapter H.2 (1 Kg of dry yeast in 10 liters of water/must mixure 10:1 ratio), let the yeast ferment at room temperature for about two hours. Then add 10 liters of the stuck must to the new starter and measure the Brix of this culture. Allow to ferment until the Brix decreases about 2 units and then add the starter to the stuck must.

    d. Low temperature - In addition, the temperature may be too cold (in white must) and, if this is the case, it should be temporarily raised to 14 - 18°C.

3. By-Products of Fermentation

    Besides ethyl alcohol, which is the major product of alcoholic fermentation, there are many by-products found in the wine as a result of the fermentation processes, which have important effects on the wine flavor and quality.

    Glycerol - by-product of a side reaction which ends up with glycerol, mainly at the beginning of the fermentation. Its production is greater at higher fermentation temperatures, so it is generally found at higher concentrations in red wines than in white. The concentration range in wine is 2 - 15 grams/L, with an average value of about 8 grams/L.

    In botrytised grape must, the glycerol concentration is high before fermentation as a result of the mold infection, so the wine made from this must will contain especially high glycerol concentration. Glycerol has a sweet taste (about 70% sweetness of glucose), and its high viscosity contributes to the wine's body.

    Methanol - not a product of fermentation, but a by-product of demethylation of pectins by enzymatic activity. It is more often found in red wines (through the pectins extracted from the skins) than in white. The concentrations found in wines range between 20 - 200 mgs/L with an average of 60 mgs/L in white wines, and 150 mgs/L in reds. In large amounts, methanol is poisonous with a lethal dose of 350 mgs/kg of body weight. The natural concentration in wine is much lower than any causing health hazards. In the body it is metabolized in the liver as ethyl alcohol.

    Higher Alcohols (Fusel Oil) - formed during fermentation mainly by decomposition of amino-acids. Aeration of the must during fermentation enhances their production. To mention some of the major mono-alcohols: 1-propanol, 2-methyl 1-propanol, 2-methyl 1-butanol, 3-methyl 1-butanol, 2-phenyl ethanol. Their total concentration is about 150 - 500 mgs/L. One higher di-alcohol which is found in relatively high concentrations in wine is 2,3-butanediol with an average concentration of about 500 mgs/L.

Succinic Acid - by-product of the fermentation, mainly at the beginning, and found in relatively high concentration (about 1% of the alcohol formed). Its range of concentration is 500 - 1200 mgs/L which contributes to the wine acidity.

Volatile Acids - mostly acetic acid, but there are traces of formic, propionic, and butyric acids as by-products of the fermentation. At normal fermentation, the range of acetic acid is around 300 - 400 mgs/L. It becomes a serious problem when the wine has been infected by the aerobic acetobacteria that produce acetic acid at concentrations which are not tolerable in wine (by taste and smell) above 1500 mgs/L. The legal limit of volatile acids in most countries is about 1200 - 2000 mgs/L, expressed as acetic acid. In infected wine, the acetic acid is accompanied with ethyl-acetate which is the substance that causes the unpleasant acetified smell in the spoiled wine.

Lactic Acid - D(-) lactic acid is a by-product of the fermentation with a concentration range of 100 - 1000 mgs/L. In malolactic fermentation (next section), it is the main product, with L(+) lactic acid configuration. In this case, its concentration depends on the original amount of malic acid present, which is usually in the range of 2 - 5 grams/L.

Acetaldehyde - by-product of fermentation, the oxidized form of ethanol. It is bound easily to sulfur-dioxide or to ethanol. The concentration range is 50 - 100 mgs/L. It is also the main product of flor-sherry yeast fermentation.

Hydrogen-Sulfide - produced in two possible ways: first, during fermentation, by reduction of sulfate or sulfite or even elementary sulfur (from the grape dust), especially when there is nitrogen deficiency. The other, is by decomposition of dead yeast cells in the yeast lees if not racked off in time. Its existence is noticeable at very low concentrations in the range of micrograms/L.

In the alcoholic wine media hydrogen sulfide may react with ethanol to produce ethyl-mercaptan (ethyl-SH) and even disulfide (ethyl-S-S-ethyl), which smell much more unpleasant than hydrogen-sulfide itself and are much more difficult to get rid of.

4. Malolactic (ML) Fermentation

In the malolactic fermentation, L(-)malic acid (four-carbon backbone, di-acid) is transformed into L(+) lactic acid (three-carbon mono-acid). The fourth carbon is released as carbondioxide:

$$COOH\text{-}CH(OH)\text{-}CH_2\text{-}COOH \text{ ---> } COOH\text{-}CH(OH)\text{-}CH_3 + CO_2 \quad (13)$$

This reaction is carried on by the malolactic bacteria if the right conditions are met. One of the conditions is temperatures above 17 - $20°C$. Years ago, in Europe, when spring came and suddenly the dry wine began to "referment", it was said that the wine "wants" to return to its source - to the vineyard. This "refermentation" was the ML fermentation. Whether ML fermentation is desired or not depends on the case at hand. The main results of this fermentation are reduction in the total titratable acidity (di-acid become mono-acid) and development of a special "vino" bouquet. In white wine, the fruity flavor is reduced, so generally this fermentation is not desired. In red wines, it adds complexity and enhances wine bouquet so it is generally most desirable.

One of the major by-products of ML fermentation is diacetyl ($CH_3\text{-}CO\text{-}CO\text{-}CH_3$) which at low concentrations contributes to the wine's bouquet, where at higher concentrations it has an off-flavor.

The ML bacteria also consumes citric acid and transforms it to acetic acid, which is distinctly unpleasant. So for wines (or must) which are intended to have ML fermentation, it is advisable not to add citric acid when acidity correction is made. If malic acid is added for this purpose, the commercial compound contains the two isomers, D and L, in equal amounts. The natural one in grapes is the L, and this is the only isomer which the ML bacteria will consume, leaving the D-isomer untouched.

The ML fermentation is a microbiological instability factor which must be taken into account during wine processing. Either the wine should proceed with ML fermentation and become stable in this respect, or measures should be taken to prevent it. To inhibit ML fermentation, the following measures may be taken:

a. **Early racking** after yeast fermentation which reduces growth nutrient for the ML fermentation and also removes a great portion of the ML bacteria present in the wine.

b. **Early fining** and filtering for the above reasons.

c. **Sulfur-dioxide** should be added to maintain 25 ppm of free SO_2. The ML bacteria is very sensitive to SO_2 which can completely inhibit its growth.

d. **The pH** should be lower than 3.3 for red wine and 3.1 for whites. At low pH the ML growth is almost completely inhibited. In high acidity wines, when ML fermentation is necessary to reduce its acidity, the major problem is that these wines are also low in pH, which makes the ML growth very difficult. In such cases it is necessary to partially deacidify the wine and bring its pH to the minimum level which enables the ML growth.

e. Keeping and storing the wine in the cellar, up to bottling, at temperatures below 14°C. The temperature inhibition factor is quite effective, with some special exceptions at temperatures as low as 10°C.

f. At bottling, membrane filtering at 0.45 micron size should be employed to assure that ML fermentation will not start in the bottle.

g. Chemical inhibition, which is very effective, by fumaric acid at concentration of 300 - 500 mg/L (see chapter G.1).

If all or part of these measures are practiced, there is a very good chance that ML fermentation will not take place.

In order to encourage ML fermentation, all the opposite measures should be taken. In red wine, where ML is generally desired and if must acidity is too low, the necessary acid correction should be made only after ML fermentation is completed.

The ML bacteria which grow in wine belongs to these three genus: *Lactobacillus, Pediococcus* and *Leuconostoc*.

The first two genus are undesirable in wine. They can cause off-flavor and unpleasant after-taste. If sorbate had been added to the wine to prevent some residual sugar from being fermented (chapter K), the ML bacteria may transform it to 2-ethoxyhexa-3,5-diene which has a "geranium like" off-flavor.

The pure ML culture which is used in winemaking is the *Leuconostoc-oenos* species, with some different strains such as ML-34, PSU-1 and 44-40. This species are round-shaped bacteria about 1 micron in diameter.

To assure ML fermentation when it is desired, the best time to add the ML starter is a few days after the yeast fermentation starts, while the alcohol concentration is still low. The yeast fermentation and the ML fermentation will proceed simultaneously. After the yeast fermentation comes to an end, the wine should be left on the lees for about a week until the ML fermentation is completed and then racked. Sulfur dioxide can be added (although it is better to avoid it) to the must at crushing, but not more than 20 - 30 ppm. Addition of more SO_2 can be done only after the ML fermentation is over. In Germany, for example it is not allowed to inoculate with ML culture. Generally in German wines it is not desired by the style, and if it is desired, it should start by itself which is very difficult because of the low pH and low environmental temperature.

To prepare the ML starter for inoculation, follow these general guidelines. At crushing, take about 3% of the must volume and dilute it with 10% water. Do not add sulfur dioxide, and if the pH is lower than 3.5, bring it to pH 3.5-4.0 by using calcium carbonate or potassium carbonate. Add 5 grams/L yeast extract, 20 mg/L of dry yeast and 20 ml/L of commercial ML culture. Shake well, close with a fermentation lock and place in a warm temperature, 25 - 35°C. In 3 to 5 days the starter may be ready. The number of bacteria cells, when ready, is about one million/ml for good inoculation. Add about two-thirds of the starter volume to the main lot, when the yeast fermentation is in the exponential growth stage (a few days after starting). Follow the malic acid disappearance (and lactic acid appearance) by use of paper chromatography (see chapter R.9). The third volume of the ML starter, which was left over, can serve as the inoculation source for a new starter batch by addition of a new must. After ML fermentation the fixed acidity has been reduced and if acid correction has to be made, it is recommended to wait until cold-stabilization has taken place, and then the exact correction can be made.

K. RESIDUAL SUGAR

1. Classification

The term residual sugar refers to any concentration of sugar left after the alcoholic fermentation is over. It may vary from 2-3 gram/L in very dry table wine, up to 100 - 200 gram/L in late harvest botrytised dessert wine.

In good quality wine, no addition of sugar is professionally accepted. The sweetness should come only from the natural grape sugar. In certain dessert wines, the addition of grape juice or grape concentrate is common practice, is legal and well recognized.

There are many types of wines containing sugar on the market. They can be clasified into three major categories :

a. Off-dry table wines with about 1 - 3% sugar. In certain table wines, especially if they have fruity and fresh character (White Riesling, Chenin-Blanc, French-Colombard) it is best balanced with some sweetness. In Muscat related varieties (Muscat of Alexandria, Muscat Canelli, Gewurztraminer, Sylvaner, Symphony), it is almost essential to leave some residual sugar in order to mask the bitter after-taste characteristic of these varieties. Also, residual sweetness is in good harmony with the very spicy and powerful perfumed aroma of these varieties. Most of the "blush" wines ("blanc de noir") also belong in this category.

Almost any fermentation, when going smoothly will leave some residual sugar (r.s.) at about 0.2 - 0.3%. This residual sugar is mainly unfermentable pentose sugars, arabinose and rhamnose. There may also be some unfermented fructose and glucose. If the fermentation is stopped before all fermentable hexose sugars have been consumed, more fructose will be found than glucose, because the glucose rate of fermentation is slightly higher than that of fructose. Besides sugar, there is another sweet component in wine, glycerol, at a concentration range of 0.2 - 1.5% with about 70% sweetness of glucose. Ethyl alcohol also has some slightly sweet taste, so even when the fermentation has come to "complete" dryness, some people experience a sweet taste in dry table wine, especially whites.

b. Sweet dessert wines such as: Port, Sherry (Oloroso), Madeira, Marsala and others. Each wine in this category has its special character depending on its unique version of production. The residual sugar content in these wines is usually between 5-15%.

c. Late harvest production with two main styles:

a. The German late harvest wines (Auslese, Beerenauslese, Eiswein) which are made of over-ripe or shrivelled grapes where the sugar level became very high in these almost dry grapes. The most popular varieties in this version are White Riesling and Gewurztraminer. Hungarian Tokay can also be classified here.

b. The Sauterne (Bordeaux) wines which are made of late harvest grapes infected by *Botrytis Cinerea* mold ("Noble Rot"). The infection, which usually causes the grapes to become rotten and undesirable as table wine, may in certain weather conditions (high humidity at infection time, and dry and warm in the following days) develop in such a way that they cause the berries to lose their water without rotting. The sugar level (and all dry extract) becomes very high to about 30 - 50 Brix. The Botrytis not only concentrates the dry extract in the berries but also gives some very special aroma characteristics to that style. The wine made from these grapes ends up with about 8 - 20% residual sugar. The most common varieties for this type of wine are Sauvignon Blanc and Semillon, but White Riesling or Gewurztraminer are also suitable.

When such wine has to be made, its processing is quite different from table wine. The right times to pick the grapes is most crucial. If picked too early, the sugar will not be concentrated enough, and if picked too late, the grapes might be totally rotten. Several pickings may be necessary according to the bunch "Noble Rot" development and even picking berries individually, is practiced in some wineries.

The pressing of the shrivelled berries (whole clusters, no crushing/destemming), needs high pressure for a very long time, and many cycles of pressing. The juice contains a very high percentage of solid lees (which will be difficult to clear up), and a high content of aldehydes and glycerol. No settling is done and the general practice is to add about 100 ppm SO_2 before fermentation, which will probably be fixed by the aldehydes. Diammonium-phosphate (DAP) is also recommended, (200 ppm), to ease the fermentation. The inoculation is best done with yeast culture rather than with dry yeast, in order to acclimate the culture gradually to the very high sugar level of the must. It is also preferable to start the fermentation at a moderate temperature (18 - 20°C) and only after it has developed well, reduce the temperature to 10 - 14°C. After fermentation has stopped at a certain sugar level and the wine has been racked off, fined and stabilized, it can be aged in oak barrels for 6 - 18 months before bottling. In certain places the must itself is fermented in oak barrels.

Addition of SO_2 is necessary to maintain about 20 - 30 ppm of free SO_2 during all the processing stages up to bottling. Because of the initial high SO_2, the total SO_2 in late harvest wines is usually considerably higher than in regular table wines. 150 - 250 ppm is a common level in late harvest wines.

It must be mentioned that the grapes in the German late harvest version may sometimes also be infected with Botrytis.

2. Stopping fermentation and preservation

There are several techniques to stop the fermentation before dryness and to preserve it from refermenting. Stopping fermentation is the "natural" method to leave a certain amount of residual sugar in the finished wine. It can be done in fully sweet, semi-sweet dessert wines, or off-dry table wines.

i. Deep Cooling

Most suitable for premium table wines. If this technique has been chosen, right from the beginning the winemaker should make the yeast's condition not very comfortable; under stress the yeast will be weak enough to be easily controlled. No nutrition should be added before fermentation (lees settling is obvious and using bentonite on the must prior to fermentation is also possible). Yeast inoculation can be done with half the regular amount, and low fermentation temperature 10 - 12°C is the main regulation factor.

Although any yeast strain can be used, the yeast strain which is easiest to stop by cold-shock is *Epernay-2*. This strain is also good for fruity and highly aromatic white wines. Under all these conditions, the fermentation may last longer than usual, 2 to 4 weeks. Toward the end of the fermentation, the alcohol and sugar analysis should be made and at the desired sugar concentration, the temperature should be drastically lowered to about 0°C (32°F). The fermentation will terminate. If the fermentation is stopped at a high sugar level, while the yeast is very active and the rate of fermentation is still high - there may be a delay until the final termination. If a strong must chiller is not available to drastically cool the must, it will take some time for the temperature to set enough just by reducing the temperature control. In such a case, the decision to stop the fermentation should be made at the extrapolated Brix reading which is higher than the desired residual sugar level. For such extrapolation, a daily record of the Brix level is necessary.

On the other hand, if the fermentation is halted close to its end, when the rate naturally slows down, such measures are less important because the sugar concentration changes are slow. In about a week or two, the wine has to be racked from the yeast lees, bentonited, and cold-stabilized. After cold stabilization, the wine should be filtered (pad filter 30 -50) and aged in a tank until bottling, at temperatures not exceeding 5 - 8°C.

ii. Natural Stopping

In some cases, just by imposing stress on the yeast, the fermentation may stop by itself. The stopping is unpredicted with no control on the desired residual sugar level in the wine.

In late harvest must the conditions are such that the fermentation is always under stress. The nutrition is usually too low, and the high sugar concentration is an inhibition factor for the fermentation, especially when the alcohol becomes high enough (around 8% - 12% alcohol, depending on the residual sugar concentration). So, in most cases the high Brix botrytised late harvest must, will stop its fermentation naturally.

iii. Alcohol Fortification

Another technique to stop fermentation and leave residual sugar, is by addition of alcohol to the fermenting must up to final concentration of 18% (by volume), where the yeast cannot tolerate anymore and dies. The addition of alcohol is done at the desired sugar concentration, usually at 6 - 12% in dessert wine production, like Port, Sherry, Madeira and others. After the addition of alcohol, the wine is well mixed, racked from the yeast lees after a few days and cold-stabilized. The high alcohol and sugar concentration will keep the wine from refermenting during the rest of the wine processing and in the bottle.

There is a different technique to provide residual sugar in wine, which is less troublesome and less caution needs to be taken.

This is most suitable for regular less expensive off-dry table wines. In this technique the wine is fermented to dryness, racked from the yeast and then bentonited. After bentonite racking, a calculated amount of sugar is added as grape juice concentrate, to bring the wine to the desired 1 - 3% sugar level. Stabilizing and aging is carried on as any regular wine, but until bottling the wine should be kept at a temperature of less than 5 - 8°C to prevent refermentation. Another way to sweeten the wine is by adding "sweet reserve" (unfermented grape juice) which has been kept in the cellar since harvest. After lees settling the clear must is sulfited up to 400 - 500 ppm SO_2, and stored in a tank, at 0 - 3°C.

Less care need be taken if the sugar addition is made just before bottling, where the addition is followed by microfiltration. The only risk with such late sugar addition is that it may change slightly the wine's stability in the bottle. However, if the wine is to be consumed young and early, this risk is minimal.

Any fermentable sugar is a major microbiological instability factor in the wine and special attention has to be paid to its presence. The last section of this chapter deals in ways of how to preserve wine from refermentation :

i. Yeast Inhibitors

The most common agent to inhibit fermentation is sorbic acid or its salts. Before bottling, about 200 - 250 ppm of potassium- sorbate can be added. The sorbate has an inhibitory effect on yeast growth, and should be used in no less than 200 ppm in order to be effective. In complementary action with the sorbate, 30 - 40 ppm of free sulfur-dioxide is most recommended. The sorbate has almost no offensive taste or odor at these levels, especially in sweet wines. Its taste threshold is around 150 ppm, and its legal limit in the United States is 1,000 ppm.

The main objection to sorbate is the 'geranium like' odor which may develop by lactic bacteria from sorbic acid to produce 2-ethoxyhexa-3,5-diene. To some people this compound may be pleasant, to others unpleasant. The SO_2 presence with sorbate is a good assurance that ML-bacteria, will not be active to carry on that reaction. The sorbate is widely used in sweet wines which have not been fortified or pasteurized.

ii. Pasteurization

A very old and efficient technique to prevent refermentation of sweet or semi-sweet wines. It is not recommended for quality table wine because the heat might destroy its freshness and fruity aroma. It also may contribute a caramel taste to the wine. The lowest damage is caused by flash pasteurization at about 80°C for a few seconds and immediately cooling. After the pasteurization, the wine may become cloudy because of protein precipitation, so fining with bentonite and filtration is necessary. However, by doing so, the wine may become unsterile again from the tanks and equipment, so another pre-bottling pasteurization is needed. To avoid it, the wine can be bottled directly after the bentonite racking through pad filtering and microfiltering to the bottling line.

iii. Sterile filtration

This technique is based on mechanical removing of the yeast from the wine, and keeping it under such conditions which make fermentation very difficult to restart (low nutrient, alcohol, SO_2, and very very small yeast population). Up to bottling, when sterile filtering is done, the wine has to be checked regularly for signs of refermentation (CO_2 bubbles, microscopic test).

At bottling, the sterile filtration (0.45 micron), does not allow any wine spoilage microorganism to get through. In the modern technology of winemaking, this is the most used technique to prevent refermentation in the bottle. For details see chapter 0.2.

L. CELLAR OPERATION

1. Racking

Wine is racked off a couple of times during processing. Racking is another word for decanting, meaning transfer of liquid above its solid sediments. In wine processing, the solids are called lees and may contain yeast cells, pulp, tartarate crystals, fining particles, proteins and tannins.

In the racking operation, two major changes occur: clarification and aeration.

The main principle in racking is simple: as little as possible. This means that the racking should follow only certain operations, which leave sediment at the bottom of the tank which must be separated from the wine.

The operation of racking is done this way: a hose is connected to the racking valve in the tank, which has an inner bent tube inside the tank that can alter the level of suction. Before this tank is filled, the angle of this inner tube should be at its highest inlet which is just below the cleaning opening level of the tank (see figure L.1). The

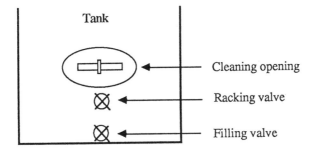

Fig. L.1 - Tank's openings.

procedure is to pump the wine out of the tank, up to the outlet level and then open the cleaning opening and check the surface level. By gradually lowering the inner tube down (from the opening), it is possible to pump out the wine exactly to the lees level. In tanks without the inner tube, the procedure is a little different. Again the hose is connected to the racking valve and the wine is pumped out until the liquid surface reaches the racking valve level and the pumping is stopped. Then the cleaning opening of the tank is opened, the hose is disconnected from the racking valve and placed over a cellar worker's shoulder and through the tank's opening into the liquid surface. The wine is then pumped out carefully up to the lees layers. The other side of the hose is connected to the receiving tank, either at the bottom into the filling valve if no aeration is desired, or at the top of the tank when aeration is desired. All connections between tanks, hoses and the pump should be sealed well, with no leakage of liquid or leaking in of air.

The racking operation can also be used to administer and mix the necessary materials during wine processing such as fining agents, sulfur dioxide, acids, or blending with another wine. When this is done, after finishing the racking, in order to complete the mixing, the wine can be circulated in the tank from bottom to top.

Any aeration during racking introduces oxygen into the wine at concentrations from tenths of mg/L in careful racking, up to 5-6 mg/L of oxygen in over-aerating racking. For white wine, less oxygen is better, to prevent browning and preserve freshness. In red wine, certain amounts of oxygen are necessary during the process of aging, so racking with some aeration is good practice.

In order to minimize oxygen absorption during racking, the receiving tank can be filled prior to racking with CO_2 gas through the bottom valve, for a couple of minutes. The CO_2 which is heavier than air, stay on the wine surface and will protect the incoming wine when it fills the tank.

The major rackings are:

First Racking - after fermentation has terminated. The lees are mostly yeast cells and pulp. This racking should be done shortly after fermentation stops, to prevent extraction of dead yeast constituents into the wine, like protein metabolites, amino acids and hydrogen sulfide. In red wine, there is a large quantity of seeds in the lees and if left unracked for a long time, the wine will extract much of the tannins.

In certain styles of wines, whites or reds, the first racking is delayed for a long time, regardless of the above considerations. In white wines it is done especially in selected lots of Chardonnay which are barrel fermented and left over after the fermentation for a couple of months on the lees. These lots are later blended with the other regular lots.

In some places in Europe, red wine is left over the yeast lees for several weeks before racking.

Second Racking - after fining (with bentonite) and cold-stabilizing. The lees consists mostly of fining particles and tartarate. In whites, the racked wine is stored in tanks until bottling. In red wine, it is generally racked into barrels for aging.

Third Racking - in white wine, from the tanks to the bottling line through filters. In red, from the barrels to unify the wine and blend it in a tank for further storing and aging. During the barrel aging of red wine, racking from barrels to other barrels is done every three to six months. The empty barrels (after racking off) are washed with water and refilled from another racked barrel.

Fourth Racking - red wine is racked from the tanks to bottling through filters.

This racking scheme presents a general framework. In some cases, because of certain reasons, some lots must be treated and racked more than the above description.

2. Stabilization

This term refers to operations which prevent cloudiness and settling of particles in the bottle. The causes of cloudiness are protein coagulation, polyphenolic colloids precipitation, metals haze (iron and copper) and tartarate crystallization. Protein and polyphenolic stabilization will be discussed in the fining chapter (M). Iron and copper haze which is very rare now, as all cellar equipment is made of stainless steel, will be discussed in the quality control chapter (P.2).

In this chapter we shall deal with tartarate stabilization. Tartaric acid is very soluble in wine. Its mono-salt, potassium- bitartarate which is less soluble, precipitates out during wine processing. The reason for its precipitation is the presence of alcohol in the wine, an effect known as salting-out. In this effect, the less-dielectric solvent (alcohol) added to water, reduces the solubility product of ionic substances causing them to precipitate if their concentration is close to their saturation point. A good example of this effect can be seen in figure L.2 which shows the solubility of potassium-bitartarate in

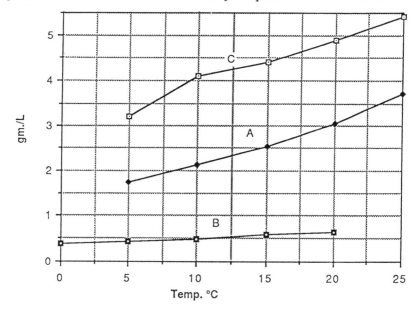

Figure L.2 - The solubility of potassium-bitartarate (line A) and calcium-tartarate (line B) at different temperatures in water containing 10% (v/v) alcohol. Line C is the solubility of potassium-bitartarate in pure water.

water and in water containing 10% alcohol, at different temperatures. Because the solubility is temperature dependent, and potassium-bitartarate in wine is generally close to its saturation point - it will precipitate when the temperature lowers.

The purpose of potassium-bitartarate stabilization is to prevent crystallization of the tartarate in the bottle, by removing the excess amounts of it. There is nothing wrong in tartarate crystals ("cream de tartar") in the bottle, from the quality point of view; and in Europe the customer is more tolerant of it. In the United States it is considered an aesthetic defect to have any kind of precipitation in the bottle, regardless of what it is.

The factors which determine the solubility of tartrate in wine, besides temperature, are alcohol concentration, potassium concentration, the pH, other anions and other cations like iron, magnesium and calcium. The stabilization is brought about by lowering the temperature to $-2°C \rightarrow -5°C$, which is slightly above the freezing point of wine (about $-6°C$).

There are some formulas and tests which can be done to determine the optimal temperature for stabilization. We think this is of no importance, and that simply by keeping the wine at the range below $0°C$ and above the freezing point ($-5°C$), for two weeks, the potassium-bitartarate stabilization will be set. The cooling can be done either by passing the wine through a heat exchanger (chiller), resulting in quick cooling to the desired range, or by lowering the temperature of the tank by its temperature control, which takes a couple of days. In our opinion, the latter tactic is preferred, especially in white wine, because it saves the need for one more wine transfer. To decrease supersaturation and to ease the crystallization and precipitation, it is recommended to combine the bentonite treatment (chapter M.1) with the stabilization operation. In such a combination, the bentonite is added with the right amount of sulfur dioxide (25 - 35 ppm), mixed well and then the temperature is gradually reduced by lowering the temperature control. The fining particles will facilitate the tartarate settling. After two weeks, the wine is racked and filtered cold for aging before bottling. The winemaker saves one racking, in comparison to when these two operations (fining and stabilization) are done separately.

It is recommended to check the total acidity (and better the tartaric level, if possible) before stabilizing and during cooling. By following these changes, one can know when settling has come to equilibrium.

After cold stabilization, there will be no more potassium-bitartarate precipitation. However, it should be mentioned that sometimes, though quite rarely, there may be a calcium-tartarate precipitation even after stabilization. This can happen when the natural L-tartaric acid isomer is gradually isomerized to a racemic one (very slowly). Because the solubility product of racemic calcium tartarate is eight times smaller than L-calcium-tartarate, if there is enough calcium ion concentration, it will precipitate. In white wine, if consumed young within 1 to 2 years, there should not be a problem. In red wine the tartarate is more stable in solution because of the tannins and, in any case, some sediments in red wines after many years in the bottle are tolerable and accepted.

The last thing to remember is that after blending wines, there might be an instability potential and a cold stability test is recommended (see section on blending).

3. Filtering

Cloudy wine will clarify if let stand for a long time. However, even a wine which looks clear after a long time of settling, contains enormous amounts of particles in the microns (10^{-3}mm size) range, including yeast cells, bacteria, pigment particles, proteins, fining particles, tartarate crystals, pulp and other components. The purpose of filtering is to facilitate the removal of all these particles from the wine.

When wine flows through a filter, the rate of filtration (volume filtered per time unit) is proportional directly to the pressure difference between the input and output of the filter, the filter area, and inversly proportional to the viscosity of the liquid, and to a factor called "resistance to filtration". This depends on two sub-factors: the filter's resistance to clear water flow, an inherent characteristic of the filter texture (pore size and packaging tightness), and on the liquid filtrate materials (particle's size and shape and their number per volume unit). This dependence can be expressed in a formal relationship :

$$\frac{dV}{dt} = \frac{dP \times A}{g \times R} \qquad (14)$$

where: dV/dt - is the volume of filtered wine per unit time (filtering rate).

dP - is the pressure difference across the filter.

A - is the total filter area (pad's area x number of pads, or cassette area).

g - wine viscosity.

R - filtering resistance.

Filtration is done through two basic mechanisms ; first by absorption of the particles on and in the surface of the filter texture, by electrical or cohesion forces. These filters are made of fiber pads or a composite of fibers and mineral particles. The wine particles are trapped in these filters on the whole volume of the filter surface area. The second mechanism is by size control of the pores which prevent any particle bigger than the filter pores to get through. These filters are called membrane filters and are used mainly for very fine and sterile filtration.

i. Pad Filters - the absorption filters are made of fabric or paper pads which are mounted tightly in a stainless steel or plastic frame (see figure L.3). The mechanical set which is called a plate

filter frame (metal or plastic)

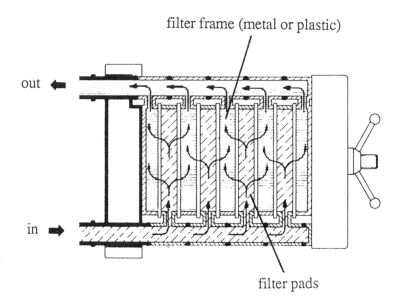

out

in

filter pads

Fig. L.3 - Frame-filter device and the wine path through it. The wine
flows through the frames leaving the solid particles inside
and on the pads.

filter, contains a series of plates that hold the pads in between (one
plate, one pad). The common dimensions of the commercial pads are
squares of 20x20 or 40x40 centimeters. The pads are not identical on
both sides. One side is rough, facing the inlet wine flow, and the other
side is smooth, facing the outlet wine flow. When the pads are
parallel packaged in the plates, they are placed alternately between
them. The flowing of the wine through the plate filter is designed so
that the pads should be placed this way. So, in placing the pads in the
plate frame, one should be aware of maintaining the proper order of
the pads, to avoid having tiny particles and fibers of the pads enter
the wine. The commercial pad filters are numbered from 20 to 70
(sometimes from 2 to 7) according to the fineness of the filter. The
higher the number, the finer the filter. The sterile filters in this group
are marked by letters (EK, EKS).

The fineness of the filter is responsible for two parameters: the
distribution of the particle sizes that can pass through the filter, and
the resistance of the filter to flow. The smaller the range of particle

sizes that can pass through, the greater the resistance. The particle sizes in wine which have to be filtered out range between sub-micron size (bacteria), up to tens microns (lees and fining particles, tartarate crystals).

The flow rate of a typical commercial filters are given in the table:

Filter mark	2	3	4	5	6	6	EK	EKS
Liter/hr/m²	6600	6000	4600	3750	3400	2500	1100	750

Table L.1 - Flow rate of water through filter pad at pressure difference of 10 p.s.i.
The flow rates are in liters per hour, per one square meter area of the pads (the area of one regular size pad of 40x40 cm is equal to 0.16 m²).

Asbestos-free filters are recommended, due to health hazards. After the plate filter is set and the pump has been connected, it is very advisable to flush the whole filter setting with 1% citric acid for a couple of minutes (in close cycle) and then with clear water in order to avoid introducing the "papery" flavor of the filter pads into the wine.

During filtering, a pressure is built up across the filter surface (between the input and output of the filter). This differential pressure should not be higher then 3 bars (about 40 psi). When it reaches that pressure, it is recommended that the filter pads be replaced.

Another important factor which must be taken into consideration is the aeration of the wine during filtration. To minimize oxidation, especially in white wine, it is highly recommended to add sulfur dioxide to the wine before filtration takes place. In the first filtration, after bentonite and cold stabilization, where sulfur dioxide cannot be added just before filtering (in order not to disturb the lees), the SO_2 should be added before the bentonite and cold stabilization take place (see the previous section on stabilization).

When the second filtration is done (bottling filtering), the sulfur dioxide may be added to the wine a couple of days before filtering.

After cold stabilization, the filter to use depends on the turbidness of the wine. Generally filter marks 3 to 5 allow a moderate rate of flow and will clear the wine fairly well. If after a short time, the filter is clogged (and the pressure gets too high), the filter should be replaced by a smaller filter mark.

Some wineries use filter aids (diatomaceous-earth or kieselguhr) which are very small particles of minerals with a very large surface area, to prevent clogging the filter pads. The powder of the diatomaceous-earth (D.E.) is mixed with the wine just before filtering and when flowing through the filter, it builds up layers on the pads containing the D.E. and the wine particles. The multilayers of D.E. eases the flow rate. The amount of filter-aid which is used for that purpose, is in the range of 50 - 100 grams/HL, depending on the concentration of the particles in the wine. In certain filter devices, the filter-aid can be administered into the wine by a dosing unit mounting at the input of the filter. But in order to use the filter-aids in filtration, special frames designed for that use should be available. The regular pad frames are not good for use with filter-aids.

ii. Membrane Filters - These filters are made of synthetic polymers (cellulose esters) which have uniform tiny holes at the micron range. The membrane filter is a surface filter, in contrast to the pad filter which is a depth filter. In the membrane filter, the particles are blocked on the surface by the hole's size, which specify the filter characteristic. The common sizes in use are 1.2, 0.65 and 0.45 microns. The bigger size will block most of the yeast cells and the finest size will block most spoilage bacteria. The membrane filter is built as a plastic cartridge mounted in a vertical stainless steel tube. The wine flows from the outside into the inside of the cartridge, leaving the microscopical particles on the outer surface. During operation the accumulation of particles on the surface increases the resistance and, causes pressure to build up. As the pressure gets higher, the flow rate falls.

At high pressure, the filter may be permanently damaged losing its submicron filtering capability. The maximum working pressure is about 3 bars, and at this pressure the filtering should be stoped and the filter be cleaned. In order to clean the membrane filter, a water flow at opposite direction (from input to output) will remove the particles which are left on the membrane's surface. The membrane filter is generally used only before bottling the wine (chapter O).

At bottling, because the sterile filtering is the major measure taken for biological stability, and in sweet or off-dry wines it is sometimes the only assurance against refermentation - it is most important to check the filter before each bottling day (and after too!). The checking is done by the so called "bubble point" test. To run the test, water is run first through the filter to fill it with water. Then a nitrogen line from a cylinder is connected to the inlet of the filter and the outlet is connected by a thin tube to an open vessel with water (see figure L.4). The main nitrogen valve is then opened and by the

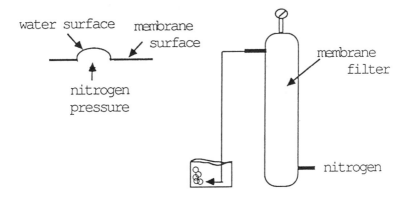

Fig. L.4 - The bubble-point test for membrane filter.

regulating valve the pressure is slowly increased till bubbles of nitrogen appear at the filter outlet (in the water). This pressure is needed to force the bubbles of gas to act against the water suface-tension across the filter's holes. The smaller the hole radius, the greater the pressure needed. If there is a break in the filter texture, the bubble pressure would be lower than the normal test pressure. For 0.45 micron cartridge (the common filter for bottling) the bubble point pressure is between 25 - 40 p.s.i. If the bubble point appears at lower pressure, the cartridge should be replaced.

iii. Mechanical Separators - There are two more devices which can be used to clear turbid wine. One is a continuous diatomaceous- earth filter (Kieselguhr) which builds a layer of small irregular particles like a filter texture on permanent support and preserves these filtering layers by constant flow pressure.

The advantages of the diatomaceous earth (D.E.) continuous machine are: high flow rate, very high filtering capacity especially when the solid particle concentration is high (must or young wine), and economical compared to pad sheet filters (D.E. is much less expensive than filter pads).

The disadvantages are: complicated operation, long preparation time before starting the filtering (about 1-2 hours), and operational error may lead to the necessity to start the whole preparation from the beginning.

The D.E. quantity which is needed for building the "cake" before filtering is about 1.5 - 2.0 kg/m^2 (square meter of supporting area). During filtering, the D.E. is fed into the wine by a doser from the D.E. container in the machine.

There are different particle sizes composing the D.E. suitable for the different filtrate particles. For must or young wine, Celite 535, 545 can be used. For clean wine, Celite 501, 503 are preferred. During preparations, before filtering the wine, the "cake" should be washed (in closed cycle) with 1% citric acid (to remove the "papery" taste) and then with clear water.

The initial differential pressure before starting the filtering is about 1 bar. The working pressure can rise up to 6 - 7 bars and then the filtering should be stopped, the machine has to be cleaned and be prepared from the beginning. The operation of this filter device is quite complicated and sensitive, and many wineries which have such a filter machine do not use it. The plate filter (with, or without filter-aid) is much simpler to operate.

The other machine for clarifying must or wine is the centrifuge, which can clear turbid wine easily and quite satisfactorily, but its price is high and perhaps not economical for a small scale winery.

A final remark on filtering: One should use only a centrifugal pump (continuous pressure) for the operation, rather than a piston pump or air membrane pump, which generates pulsed pressure on the filter.

4. Blending

Blending of wine is done in order to achieve different goals : overcoming certain deficiencies or defects, balancing the wine and enhancing complexity.

The general purpose of wine blending can be divided into two categories : to correct something wrong in the wine or to improve it. Blending can be done at any stage of winemaking. It can start in the vineyard when the vines are planted with mixed varieties (as it is done for example in Bordeaux with Cabernet Sauvignon related varieties), through must blending before fermentation, and at any stage during the wine processing. The correction made by blending two wines, enhances both of them; one is over endowed with certain character and the other suffers from a deficiency of it. By blending them, both wines will improve in that character. However, on the other hand, there are many other parameters involved in the wine quality and all of them have to be considered, rendering the "art" of blending not a simple one.

Blending can be done with different varieties, different vintages of the same variety, different vineyards or locations of the same variety, and different lots of the same vintage (tanks or barrels).

Each country has its special rules concerning the blending and labeling of wines. In the United States for example, in order to state on the label a varietal name, it is mandatory to blend at least 75% of that specific variety in the wine. If the blending is such that there is less than 75% of any variety, the wine has to be labeled as generic. This does not, however mean that a generic wine cannot be an excellent wine.

As for the region of origin (appellation), the approved American Viticultural Appellation requires that 85% of the grapes must come from the indicated region in order to be stated on the label. The regions do not necessarily coincide with geographic boundaries.

For vintage statement, 95% of the wine must be from the year that is printed on the label.

The alcohol content in table wine should be in the range of 7 - 14%.

All these rules (in the United States in this case), should be remembered by the winemaker when blending his wine. The same principles of specific regulations are practiced in other countries as well.

In varietal blending there are certain conventions regarding which varieties are suitable to blend. For example, Cabernet Franc and Merlot are commonly blended with Cabernet Sauvignon, or Semillon with Sauvignon Blanc. French Colombard is a good blend with any fruity white wine for its high acidity content and floral aroma. In certain cases of generic wines, there may be up to four varietal blends with excellent balance and quality.

In regular cellar operations, it happens quite frequently that certain lots of the same vintage develop differently, better or worse than the average of that vintage. The question that the winemaker is faced with, is whether to separate these lots (if better, to develop it as special reserve, or if worse to sell it as bulk), or to blend them together to get the best results he can, professionally and economically. No formula or recommendation in such a case can be made. The decisions are Ad-Hoc. Also if press run is left separately, at blending, special attention has to be paid to such a lot, because of its high concentration of tannins, color materials, high pH and volatile acidity.

The parameters that can be corrected by blending are : acidity, pH, alcohol, color, tannin, varietal aroma, freshness and fruitiness, oak flavor, volatile acidity, residual sugar, bitterness and off-flavor. Some of these parameters are in excess or in deficiency in the wine.

The blending may "correct" that parameter to the desired level, or at least make it tolerable.

Some of these parameters are quantitatively measurable (like acid, residual sugar, alcohol, volatile acid, color) and their concentration is linearly related. Other parameters, such as aroma, flavor, off-flavor can be appreciated mainly by organoleptic judgement.

For the measurable parameters, the volume ratios of the wines to be blended, according to the desired concentrations of the parameter in question, is given by the simple formula:

$$P_1 + P_2 X = P_b (1 + X) \qquad (15)$$

where: P_1 - concentration of the parameter in wine No.1

P_2 - concentration of the parameter in wine No.2

P_b - concentration of the parameter in the blend wine.

X - volume fraction of wine No.2 to the volume of wine No.1 (which is set to 1.0) in the blend.

For example, wine No.1 with residual sugar of 0.8% is blended with wine No.2 which has 3.8% sugar, to obtain a designed blend with 1.8% residual sugar. Then $0.8 + 3.8 X = 1.8 (1 + X)$, with a solution of $X = 0.5$. Thus, by taking 1 volume of wine No.1 and 0.5 volume of wine No.2, the blend will have 1.8% residual sugar.

For those parameters which are not directly measurable, a process of trial and error must be carried on to find the best result. All blending tests must be tried first in the laboratory, and the seeking of good advice along with close consultation with colleagues are highly recommended.

When the attempted formula has been concluded, a small sample of five gallons has to be blended and left alone for what the French call "to get married". This "marriage" which takes a couple of weeks, is then tasted and if satisfying, the whole batch of wine can be blended accordingly. If there is some doubt about the exact volume ratio, the five-gallon test can be repeated with some small variations, and after the "marriage", a decision can be made as to which is the best one.

After blending the wines at an advanced stage of processing (e.g. before bottling), sometimes a stability problem arises, especially regarding precipitation. For this reason, it seems better, that the blending is done very early, before stabilization takes place. On the other hand, only at a mature stage of the wine can one evaluate the full character and quality of the wine. This is also our opinion, and in such a case, after the blending is completed, it is recommended to leave the wine at cold temperature 5 - 10°C for a couple of weeks and to check for instability. If there is any, for whatever reason, care should be taken before bottling.

5. Maintenance

a. Sanitation

Besides the aesthetic aspect, winery sanitation is most important in preventing and minimizing bacterial and fungal spoilage of the wine. Although wine is less susceptible to harmful spoilage than other food products, it is still sensitive to a number of microorganism infections, which may affect its quality considerably. Winery sanitation includes the cleaning and sanitizing of all the equipment in the winery which comes in contact with the wine, from crushing to bottling. Of the many materials and cleaners on the market, we choose to emphasize here those which are most efficient and most used in the wine industry.

i. Water - can be used as cold, hot and steam water. Cold water cleans by dissolving and by physical removing of the soil from the surface. Hot water is much more effective than cold, and steam is the most effective because its heat capacity is about five times higher than boiling water.

Three major devices are in use for applying water cleaning in the winery : 1. a water gun, which uses the water system pressure; 2. sprinkler ball; and 3. hot water machine. Spraying water with a water gun or with an open hose is the most common and basic technique for equipment cleaning in the winery. At the end of any working day during harvest time, all the machines which have been used (hopper, crusher/destemmer, press, hoses, drainer, pumps, tanks) are washed with water to remove all must residue left over. The gun's high pressure is very helpful in removing the dried pulp particles. The sprinkler ball, which can be operated by pump pressure in a close cycle or by the water system pressure, is used inside containers like tanks and barrels to spray the water uniformly on the container walls. The hot water machine (to quote one of my colleagues), "is one of the most useful machines in a small winery". It is connected to the water line and can heat the water instantly up to about 80°C and eject it at very high pressure through a gun. The cleaning ability of this machine for any surface (metal, cement, wood, plastic) is most effective.

Besides cleaning, hot water can be used as a sanitizing agent for almost any microorganism. The effectiveness depends on the temperature and the time of application. At 80°C, 10 minutes would kill almost any bacteria and fungi (but not all spores).

For example the filling machine in the bottling line has to be sanitized at the beginning (and end) of every working day by passing through hot water at 80°C for 20 minutes.

ii. Alkaline solutions - The strongest agent in this category is sodium or potassium hydroxide. Its main cleaning function is to saponify oily soil on the surface and to emulsify it so it can be removed by water stream. Milder but still alkaline in its activities is sodium carbonate (soda ash). Alkaline solutions are also used to dissolve the hard layer of deposited tartarate on the inside walls of tanks and barrels or any equipment which had come in contact for long time with wine or cooled must. Alkaline solutions are used at concentration of about 0.2-0.5% in water. When the solid hydroxide is dissolved in water, it releases heat, and care should be taken whenever these agents are used. Alkaline solutions are skin irritating and most dangerous to the eyes. Glasses and rubber gloves are the minimum safety measures when working with alkaline.

The general practice in using alkaline is to prepare the solution in a small container 100 - 200 liters, and apply the solution to the equipment (such as tanks) by spraying it with sprinkler ball in a close cycle. The cycle contains the open solution container, pump and the tank, connected by hoses. When doing so, the tank's opening should be closed for safety reasons. About 20 - 30 minutes of application is usually enough to remove any sediment from the tank's walls. To clean small equipment, immerse it in the container for a couple of minutes. Barrels are better cleaned (if necessary) with the less strong alkaline soda-ash. They should be filled with the solution for 15 - 30 minutes, and then emptied and washed.

The residue of alkaline on the surface is difficult to remove by water. Therefore after any application of alkaline, the surface should be rinsed with water and then washed with 0.1% citric acid solution (which interacts with the alkaline surface residue) and lastly rinsed with clean water to wash out the acid.

iii. Chlorine solutions - Based on Sodium or Calcium Hypochlorite (NaClO, Ca(ClO)$_2$) which in solution can be in two forms: hypochlorite anion (⁻ClO) and hypochlorous acid (HClO), depending on the pH (see figure L.5).

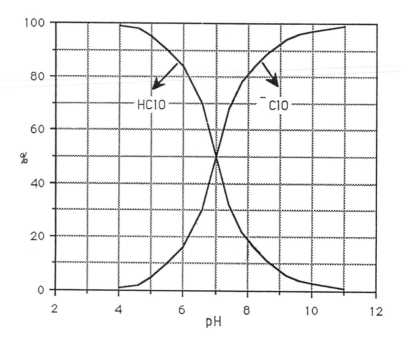

Fig. L.5 - pH dependence of the equilibrium concentration of hypochlorous acid/hypochlorite anion.

At low pH when the acid is predominate, the solution acts as an excellent sanitizing agent against most microorganisms. Its action is based on its ability to penetrate the microorganism's cell wall and chlorinating the cell's enzymes. At high pH, when the hypochlorite anion is the main form, it has a powerful oxidation potential which can bleach organic stains, cause it to emulsify and to be removed by excess water.

The commercial reagents are sold as bleaching agents (hypo-chlorite + alkaline), or as chlorinating agent (hypochlorite + acid). The first is usually sold as powder and the second as liquid.

The effective concentration is about 5 gram/L for the powder and the liquid quantity depends on its initial concentration. In any case the exact producer instructions should be followed for good cleaning effect and for safety awareness. Here again, as with the alkaline solutions, if the bleaching powder is used (alkaline), it has to be followed by washing with 0.1% citric acid and water.

iv. **Iodine** - An excellent sanitizer. It consists of solution of iodine in water (plus potassium iodine to increase solubility), and surfacetant materials to increase its surface contact. The pH is about 4 by phosphoric acid. This solution is active against microorganizms due to the free iodine. The brown color of the solution makes it easier to follow its removal with rinsing water after use. The effective concentration is about 25 mg/L, but the exact quantity should be read on the producer's instructions.

A last remark on sterilization concerns ML fermentation. Some of the wines in the winery stock are ML fermented (mainly red) and others are kept from being ML fermented. By using the same equipment (hoses, tanks, barrels, pumps, filters etc.), for both kinds, the "undesired ML" wines may be infected, although due to the very low quantities of contamination, it can develop very very slowly. To avoid such risk, it is highly recommended to mark any equipment which is used with the ML wines with a marker (such as dye or a strip of cloth) and to avoid using it with the others. Before using the marked equipment with the "undesired ML" wines, it should be first sanitized (by iodine or chlorine).

b. Storage

Wine can be stored in different kinds of containers - wood, concrete, iron, plastic and stainless steel. When concrete and iron are used, the interior of the containers is covered usually with glaze, ceramics or polymeric paint. In modern wineries, wood, concrete and iron containers are not used anymore; stainless steel has replaced them because of the ease of maintenance of stainless steel and its inert surface. Plastic containers which are cheaper than stainless steel, are rarely used in the wine industry, for no specific reason. Probably in the long run their amortization is higher than stainless steel containers. Also it is harder to clean them. Oak containers (barrels and casks) will be discussed in a separate chapter (N).

Cleanliness of the containers (or tanks) is a basic requirement in cellar operation. This means that immediately after racking, the tank should be washed, brushed if necessary, and left empty and clean for its next filling. Before any filling of a tank with wine, it is highly recommended to recheck the tank through its opening.

When storing white wine, between the different cellar operations, the main object is to minimize air contact. The reasons are twofold: preventing oxidation and acetobacteria infection. In red wine, only the last reason is of major importance. To minimize oxygen in white wine storage, it is necessary to keep the tank full as much as possible, and also to fill the space above the wine with an inert gas, such as carbon dioxide or nitrogen. The gas lays on the wine surface as an inert blanket, protecting it from coming into contact with oxygen. Carbon dioxide is heavier than air and will stay on top of the wine, but, it is highly soluble in the liquid. Nitrogen is much less soluble, but it is lighter than air and diffuses out. So, with either gas used, the top of the tank should be refilled every month. The cover or top of the tank should be closed and a water air-lock be placed on it to minimize gas diffusion and to prevent pressure differences between the tank and the outside. The water in the air lock should be replaced at any gas refilling, and it is recommended that sulfur dioxide be added to the water (1%) to keep it antiseptic.

In racking white wine, it is advisable to fill the receiving tank with carbon dioxide before transferring the wine.

As for the temperatures during storage; for white wines the preferred range is 8 to 12°C and for red wines 20 to 25°C. On barrel storage and maintenance see chapter N.

M. FINING

Fining wine has three objectives: to aid precipitation of suspended materials, to reduce color or undesirable smells, and to stabilize the wine against future cloudiness. The suspended particles in wine can be proteins, pectins, metalocolloids (iron or copper), polyphenolic polymers, and crystals of tartarate. If the particle's size is small enough and if they are electrically charged, they may not settle down by force of gravity but remain in suspension, and the wine will look cloudy.

The fining agents which are used in the wine industry are positively or negatively charged, depending on their target particles.

The principles of fining are based on two factors. The first is charge cancellation between the suspended particles and the fining agent particles, allowing the colloid suspension to agglomerate and flocculate by gravity. The second factor, which causes agglomeration, is absorbtion of the suspended particles on the surface of the fining particles so they are removed from the solution by settling. Among the many fining agents, we shall discuss here the five which are most frequently used and generally recommended.

1. Bentonite

Bentonite is an aluminium silicate clay whose microstructure consists of tiny thin plates with very high capacity to swell when it comes into contact with water. The interlameller cation within the clay plates is sodium or calcium. The preferred one for wine uses, is sodium-bentonite.

The plate's charge is negative, suitable for interaction with most wine particles (which are positively charged), especially proteins, which are mainly positively charged at the wine pH. Bentonite is used to clarify a cloudy wine and to stabilize it against protein haze. The use of bentonite in white wine is a must, otherwise the wine will become cloudy in the bottle when the temperature rises. In red wines on the other hand, there is a natural mutual coagulation of the positive protein colloids and the negative tannin particles, so the protein stabilization is of a less serious problem. Its use is mainly needed after the wine has been cleared from the yeast sediment (first or second racking). Some winemakers add bentonite to the must before fermentation to help clarify the wine faster, though it may cause some fermentation problems due to a lack of nitrogen nutrients removed by the fining. If sweet or semi-sweet wine is desired, the addition of bentonite prior to fermentation may work well.

The bentonite has to be prepared before it can be added to the wine. A 5% solution in hot water 90 - 100°C should be made, and then stirred from time to time for about one or two days before it can be used. The bentonite swells in the water and the slurry looks like a heavy, highly viscous solution.

The amount of bentonite should be minimal to fulfill its duty. An excess of bentonite reduces color and aroma, and may leave some off-flavor. It may also increase the pH. One should test in the laboratory for the right amount of bentonite to use by testing a series of wine samples with increased quantities of bentonite: 0.25, 0.5, 0.75, and 1.0 gram/liter (equal to 0.5, 1.0, 1.5, 2.0 liter of slurry/HL of wine). All samples (with an untreated one as a control) should be well mixed with the bentonite slurry and allowed to stand overnight. After settling, the samples should be decanted, filtered (1 micron filter), and checked for protein stability by heating the samples for 24 hours at 60 to 65°C. Following this, the samples are left to cool at room temperature, and are checked after three days with low intensity light for the cloudiness of each sample. The appropriate amount of bentonite will be the clear sample with the least bentonite addition.

Some varieties such as Sauvignon Blanc or Gewurztraminer will need more bentonite in order to protein stablilize. Excess use of bentonite (over 1 gr/L) may strip some of the aroma. A compromise will have to be made between prospective protein instability and reduction of the wine's aroma.

When added to the wine, bentonite acts best at moderate temperatures (15 to 25°C). The sediment volume of bentonite is about 2 - 5% of the wine volume, which may cause loss of wine. It is recommended, therefore, to treat first with bentonite and then to follow with the cold stabilization treatment with this procedure : bring the wine temperature to 15 - 20°C, and add bentonite during racking operation or mix it well by pumping over or stirring. Allow to settle for a few days and then cool the wine for cold stabilization (chapter L.2) and filter it (cold). The seeds of tartarte which are formed during cold stabilization will help to reduce the bentonite lees level to a minimum.

2. Activated carbon

This fining agent is used to reduce color intensity and to remove bad odors from wine. In white wine, it reduces the oxidized brown or yellow color, or the pink color in "blanc de noir" wine. Its activity is based mainly on absorption. In excessive amounts it may leave an off-flavor. Due to its absorption capability, carbon will also absorb aroma and flavor components, reducing their concentration in the wine, so care should be taken when carbon is used, and moreover, it should be used only if no other means to solve the problem is possible. (e.g. $CuSO_4$ can reduce the hydrogen sulfide or mercaptan very successfully instead of carbon. For more details see chapter Q.2). In many other cases its action is needed, and very helpful.

The range of usage is between 1 to 5 gr/HL for color treatment and 5 to 25 gram/HL for off-odors removal. The exact quantity required should be determined in the laboratory prior to its addition to the wine. In the laboratory test, the samples should be mixed with carbon, filtered after an hour, and checked for improvements of color or smell. When added to the wine, at temperature ranges of 15 - 25°C, the black carbon powder is administered directly to the wine and well mixed. It will settle in a couple of days. To ease settling, bentonite for protein stabilization can be added after mixing the carbon.

3. Gelatin

This fining agent is a protein, positively charged, which is used to reduce tannin (negatively charged) levels in white wines or red wines. In white, it may also reduce the bitter after-taste sometimes accompanying white and in red, it may lower the astringency of young or overly tannic wine. When used in white, it should sometimes be followed by an addition of tannin to facilitate its precipitation and to prevent protein instability due to excess gelatin. Overfining may occur when the gelatin does not completely coagulate from the wine, due to lack of tannin. When overfined, the wine has a potential protein cloudiness.

Instead of tannin addition, one can use kiesselsol (a silicate suspension of about 30% in water) for the same purpose without concern about over-dose of tannin. The kiesselsol, which is negatively charged (as tannin), is very effective in removing the excess gelatin from the wine. In red wine, the gelatin can be used without any additives because of the excessive tannins present in red wines.

As with other fining agents, laboratory tests must be carried out in order to determine the right quantity before it can be added to the wine. Gelatin solution is prepared by adding 1% of gelatine powder to hot water (90 - 100°C) and stirring well to get a jelly type solution. The quantities used in wine are in the range of 1 to 3 gram/HL of gelatin for white, and 3 to 10 gram/HL for red. With a 1% solution, it is equal in volume to 0.1 - 0.3 L/HL for white and 0.3 - 1.0 L/HL for red. The gelatin solution has to be added to the wine when hot, very slowly, and while mixing the wine. For white, the tannin or kiesselsol should be added a day before, in quantitites of 1.5 - 5 gm/HL of tannin or 10 - 25 ml/HL of 30% solution of kiesselsol. The temperature of the wine when fining with gelatin can be between 10 and 25°C. After two to three weeks, the wine can be racked and filtered.

4. Egg Whites

One of the best fining agents for red wines is egg whites. Its action is to soften the wine's astringency and to mellow its feeling with no after effects. The best timing for applying it is when the wine is at the barrel aging period.

The quantity (depending on the tannin level) is one to three egg whites per 60 gallon barrel. To prepare an egg white solution, the egg whites have to be separated from the yolks and added to salted water (to ease the solubility of the globulin). The proportions are ten eggs in one liter of water containing 1.5 gram of table salt.

After addition to the barrel and after being well mixed, the wine should be racked off in one to two weeks (not later).

5. PVPP (Poly-Vinyl-Poly-Pyrrolidone)

This is a synthetic polyamide polymer which is insoluble in water and which does not leave an off-flavor in the wine. It is used instead of gelatin in white wines to reduce tannin levels, to remove the browned oxidized polyphenols, and is also effective in preventing browning through oxidation by removing part of the polyphenyloxidase enzyme. It can also remove slight off-flavors in white wine.

The quantity used is in the range of 5 - 20 gram/HL and, as usual, the exact amount has to be determined by laboratory test. It is added as 5 - 10% slurry solution directly to the wine while mixing. A small quantity of bentonite will help it to settle more rapidly. It is most effective when used with carbon to remove undesirable odor. Both agents settle down very quickly when used together.

6. Summary

A summary of the above fining materials and conditions of use are given in table M.1.

Fining Agent	Quantity Range of Use	Quantity Range in Solution	Temperature Range °C	Use For
Bentonite	(25-100) gm/HL	(0.25-1.0) L/HL	20-25	Protein removal and stabilization
Active carbon	(15-25) gm/HL	powder	15-25	Reduce color and undesired odors
Gelatine (+ tannin or kiesselsol)	(2-5) gm/HL white wine (5-20) gm/HL red wine	(0.2-0.5) L/HL white wine (0.5-2.0) L/HL red wine (1% solution)	10-25	Reduce tannins (high polymers) and bitterness
Egg whites	one-three per barrel	10 egg whites in 1L of water + 1.5 gram table salt	15-25	Reduce tannin level in red wine
PVPP	(5-20) gm/HL	powder	10-20	Reduce monomeric browning materials

Table M.1- Some fining agents with conditions and purpose of usage.

N. AGING AND OAK BARRELS

Aging wine in barrels has been common practice for many years. The cooperage containers used to be made of different kinds of wood and were mainly for storing the wine. Nowadays, the cooperages are made either from oak specifically for aging purposes, or from redwood for storage use, although the usage of wood just for storage purposes is rarely found in modern winemaking technology.

Aging the wine in oak barrels has two goals:

a. Slow oxidation of the wine, mainly the phenolic compounds which are then partly polymerized and precipitate .

b. Adding the oak phenols into the wine, which expands the complexity of its bouquet.

As a general rule, one of the most important parameters in red wine quality is the balance between the varietal aroma, the aging bouquet and the oak character. The last two angles of this triangle are achieved by aging the wine in oak cooperage.

Wooden containers are not easy to maintain, because of leakage problems, sanitation difficulties, lack of temperature control and problems of maintenance when not in use. They are also expensive.

The oak tree used to build the cooperage is generally either European oak or American. American oak has a more oaky aroma, but less extracted phenols. Experienced wine-tasters can differentiate between wines which had been aged in American or European oak. Regardless of the quality differences or any other differences, the European barrels (especially the French) are about four times more expensive than the American.

The volume range of common cooperage is from 50 gallons (200 L) up to about 1,000 gallons. The small cooperages are called barrels, where the bigger ones are usually called *casks* or sometimes *puncheon* or *ovals*. The size of the cooperage is inversely proportional to the surface/volume ratio and therefore, to the time and rate of aging capability. The most practical size range for small winery is the 50 - 90 gallons barrels.

Oxygen is dissolved in wine during any major operation, such as racking and filtering, and also during storage, either from the top surface in a stainless steel tank, or from the head-space of a wooden barrel. The saturation level of oxygen in wine, at atmospheric pressure and temperature of 20°C, is about 6 milliliters of oxygen per liter of wine, which is about 8 mg/L. In a 50 gallon barrel, the oxygen absorption per one year, caused by a regular topping and racking, is about 25 milliliters of oxygen per liter wine. This amount is four times the saturation value of oxygen. At this stage of the wine processing, the uptake of the molecular oxygen into the wine oxidation process is conducted mainly by the iron and copper ions, which play an important catalytic intermediate between the molecular oxygen and the phenolic compounds. The phenolic compounds, when oxidized, change color from red to brown, then polymerize and precipitate. When the exposure to oxygen is too high, a direct oxidation of the alcohol into acetaldehyde may take place, which results in a temporary "flat" taste. (If so, later, in absence of excessive oxygen, the acetaldehyde interacts with the tannin, and the "flat" taste disappears).

The extracted compounds from the oak are non-flavonoid phenols, mainly gallic acid derivatives : vanillic acid and ellagic acid (see Appendix B on phenolic compounds in wine). The extracted materials also contain hydrolyzable lignin, and small sugar molecules, mainly pentoses. The total phenolic extract from a new 50 gallon barrel, in one year of wine storing, is in the range of 200 - 300 mg/L. With further use, the extract yield slows down and, as general practice shows, a barrel can be continuously used and be in good extractable condition for about 7 - 10 years.

When buying new barrels, some details regarding the origin of the wood, the shapes and styles of barrels should be known. Only the French cooperage will be discussed here because it has the highest reputation in the wine industry (and the public notion), and it is most wanted despite its high prices.

The main regions in France where the oak comes from are : *Limousin, Bourgogne, Allien, Nevers, Troncais* and *Vosges*. There are some differences in the extraction rate and barrel flavor between the regions, but other functions regarding the trees growth and the manufacturing process are also very important.

The major barrel shapes are the *Bordeaux* and the *Bourgogne* types. The Bourgogne types are usually shorter and wider than the Bordeaux ones. Each type has two styles : *Bordeaux Chateaux* and *Bordeaux Export, Bourgogne Tradition and Bourgogne Export* (see figure N.1). The export styles are more massive and heavy with

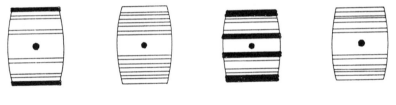

Bordeaux Chateaux   Bordeaux Export   Bourgogne Tradition   Bourgogne Export

Fig. N.1 - French barrels types and styles.

thicker staves (26 mm - 28 mm) than the Chateaux or Tradition styles (20 mm - 22 mm). This fact causes higher evaporation rate and faster aging of wine in the Chateaux style than the Export ones. The head hoops in the *Bordeaux Chateaux* style are covered with wood cut, where in the *Bourgogne Tradition* also the center hoops are covered with wood cut.

A few words regarding barrel toasting. There are three types of toasting: light, medium and heavy. The light toasting is just on the surface of the interior walls, with practically no depth.

Medium toast is more intense and penetrates to about 2 mm inside the wood surface. Heavy toasting burns the interior walls up to 3-4 mm thickness.

To compare the four options (no toasting, light, medium and heavy), it seems that the toasting serves two functions : one is to partially destroy some of the gallic and ellagic tannins, which are partly responsible for the bitterness of the oak flavor. The second function is to create a "filtering" layer of charcoal between the wine and the oak walls. Both functions cause a reduction in the oak extraction, especially in the medium and heavy toasted barrels. Attention should be paid when choosing the preferred toasting level.

When buying new barrels, the winemaker has to choose from many combined options, the ones which will suit his needs. Four of the major criteria : origin, volume, style and toast are summerized in table N.1 :

Origin	American		French Limousin, Bourgogne, Allier Nerver, Tromcais, Vosges	
Volume	30 gal.	60 gal.	90 gal	120 gal
Style	Bordeaux Chateaux, Export		Bourgogne Tradition, Export	
Toast	No Toast	Light	Medium	Heavy

Table N.1 - Barrels criteria and options to choose on buying new cooperage.

During Aging in the barrel, the liquid slowly evaporates. In the beginning the losses are quite high, but after the evaporation stabilizes, the loss is about 6 - 8 L/year. Both water and alcohol evaporates from the barrel through the wooden walls. At external humidity below 60%, water evaporates more than alcohol. At higher humidity, alcohol predominates. As a result of this evaporation, the dry-extract concentration of the wine increases, and in 3 years of barrel aging, it may gain up to 10 - 15%. This is true for the total dry-extract. As for the phenolic compounds, some of them (non-flavonols) are added from the barrel into the wine, while others (flavonols) partly polymerize and precipitate out.

The best time to put the wine in barrels diversifies widely. Before dealing with this diversity, let us clear up what kinds of wine can be aged in barrels. White wines which have a fruity character, will loose their freshness by being aged and oxidized, so there is no reason to put them in oak barrels. These wines should be ready for bottling in 3 - 6 months and be consumed young and fresh. Included in this category are varieties such as Chenin-Blanc, White Riesling, Gewurztraminer, French-Colombard and the Muscat family. White wines with a herbaceous character will generally benefit from aging in oak barrels. The varieties belonging to this category are Chardonnay, Sauvignon-Blanc, and Semillon. These wines can be aged in barrels for any length of time, from some weeks up to a couple of months.

As for reds, quality red wines have to be aged in oak barrels up to the right balance between the varietal aroma, aging bouquet and oak character, as mentioned before. The length of time may vary over a very wide range depending on the kind of wine, the variety, the amount of tannin present in the wine and the age and quality of the barrels. Generally, it can take between one to three-four years. During barrel aging the wine has to be tasted from time to time in order to decide when is the right time to rack it off from the barrel. One should keep in mind that the wine should not stay in the barrel up to its optimum maturity, where afterwards it will only decline in quality and will be overaged. The wine should be racked off from the barrels at a certain point, be bottled and left over for slow bottle-aging before reaching its full maturity.

The wine can be placed in barrels either shortly after fermentation, or later after ML-fermentation where some yeast settling has taken place.

In our opinion (and it is shown in the block-diagram), the best timing to start barrel aging (reds and whites) is after crude clarification and tartarate stabilization. At this stage of processing, when the wine is placed in the barrel, there is no need for many barrel rackings during aging, and the wine can stay in the same barrel without disturbance for a long time (3 - 6 months) until the next barrel racking. By following this practice, after each racking, the barrels are almost clean and ready for refilling without any special treatment besides washing with plain water. Roughly, the time for starting the barrel aging is about 4 - 6 months after fermentation. During the first year, the racking schedule from barrel to barrel is at intervals of 3 - 4 months. In the second year, 6 months of aging between rackings are needed, and if left for another year, no racking is needed during that year. At each racking, the sediments at the bottom of the barrels, are washed with water (high pressure sprinkler), and the barrels are refilled with the same wine. Two methods can be used for barrel rackings : one is to rack all the barrels into one big tank, then to wash the empty barrels, and to refill them back from the tank. The second method is to rack certain number of barrels to other spare barrels, to wash the racked barrels and to rack another set of barrels into the washed ones, and so on. By doing it this way, there is no blending of the whole lot of barrels, and the wine in each barrel stays separately up to the time where each barrel is tasted and judged for its special character.

No filtering of red wine is necessary and the only clarifying is done by slowly barrel settling and rackings. At the time of bottling, the wine is very clear, and only membrane filtering on the bottling line (chapter O) is needed.

Tannin fining in reds is done in the barrels with egg whites (for details see chapter M).

Some hints regarding barrel handling :

● Before transferring the wine into the barrels they have to be prepared, either by previous use with another wine, or if not, by swelling the barrels with water and checking for leakage.

● White wine should never be placed in barrels that have been used for red wine. The red pigments will color the wine even when the barrel has been well cleaned and washed. Also, good quality white wine is never placed in new barrels. Otherwise, the tannin extract will be too strong for a premium quality white wine. An elegant way to soften new barrels, is to ferment white must in it (Chardonnay or Sauvignon-Blanc). Most of the extra oak extract will be absorbed by the yeast lees, and will not affect the wine. The barrel at the same time will be less "hard" for receiving new clear wine for aging. When using such a new barrel, the wine must be checked quite frequently by tasting before it gets too "oaky".

● Do not be too concerned if the wine gets some extra oak flavor. The oak flavor decreases and becomes mellow with time, and the wine can always be blended with non-oak wine. One should never age all of the vintage wine in barrels. Leave some out of the barrel-aging process, in order to have the option of blending it with over-aged, over-oaked wine, if this becomes necessary.

● During barrel aging, the wine should be placed in shadowed, cool place, at a temperature of 10 - 20°C.

● Due to evaporation, and sometimes (especially at the beginning), due to leaking, there is a need to add wine to the barrel from time to time. This is called topping. Generally, it is done about once or twice a month, although at the beginning it should be done about once a week. It has been found that by tilting the barrel, so that the bung is at about 2 o'clock angle, the bung stopper remains wet, and blocks the entrance of air into the barrel. In fact, a vacuum may develop in the barrel by the gradual evaporation. By this tilting, topping is avoided, and can be done only when the wine is racked from barrel to barrel. The disadvantage of this practice is the impossibility of checking and tasting the wine over very long periods.

In recent years, silicone rubber stoppers were developed for barrel use. These stoppers can close the bung so tightly, that vacuum can develop. So, when using these stoppers, there is no need for tilting the barrel to maintain vacuum. Tasting the wine between the rackings is no longer a problem.

● Barrel maintenance is of very special importance, because barrels are very expensive and easily become contaminated, causing wine spoilage. New barrels should be soaked for a day with hot water a couple of times, to reduce any excess of bitter constituents. Used barrels should be soaked with a 0.1% solution of potassium hydroxide or 0.2% soda-ash, for half a day, then be soaked with 0.2% citric acid for half a day, and then washed with hot water a couple of times. After washing new or used barrels, sulfur dioxide should be put in, either as gas from an SO_2 tank, or following the old tradition, sulfur sticks can be burned in the barrel to produce the sulfur dioxide. Caution should be taken to prevent melted sulfur from falling into the barrel which later on when the barrel is filled with wine, may be transformed into hydrogen-sulfide.

● When mold grows on the outside surface of the barrel, it can be washed with an ammonium-chloride solution (3%) which will inhibit further growth. If there is mold inside the barrel (which can be verified by its smell), it is a real problem. The best treatment is first to open the barrel and scrape the interior surface, and then to soak the barrel with a hot (60°C) solution of potasium hydroxide (1 to 3 gram/L) for a couple of hours, and then with 1 gram/L solution of hypochlorite for a couple of hours, and finally rinsed with hot water a couple of times.

● When the barrels are not in use, maintaining them requires full attention. If left to dry out, the wood shrinks, the hoops loosen, and the barrel's form can deteriorate. When this happens, the barrel does not hold liquid and it takes a long period of wetting for the staves to expand again tightly into place before the barrel stops leaking. On the other hand, if the barrels are left full of water, they may develop mold and other kinds of infections. The best way to store unused barrels for long periods is to leave 1 to 2 liters of water inside, and fill the remaining space with SO_2 gas (by opening the SO_2 valve for two seconds) and close the bung. About once every two months, water and SO_2 should be added to maintain the moisture and SO_2 environment inside the barrel. For short-term storage, the barrel can be filled with a water solution of 300 ppm of potassium-metabisulfite and 100 ppm of citric acid. The barrel should be filled with that solution and checked from time to time to maintain the SO_2 level.

● It is quite common that on drying and wetting the barrel, the bung stave cracks at its weak point - the bung hole.

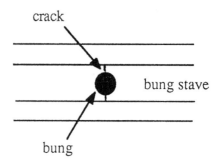

When such a crack occurs, and if it is not a serious break, it will not affect the barrel's functioning, and the barrel can be used as usual.

● The external surface of the barrel should be kept clean and dry. The hoops should be either painted against rust or be galvanized. The wood should be covered with linseed oil from time to time to protect it from wood borers. The oil has to be applied when the barrel is empty and completely dry outside.

● When not in use, the barrels should be stored under the same conditions as those which are in use (dry and clean, in a shady, cool place).

O. BOTTLING

A bottle of wine is the end product of the winery. It is the winemaker's pride, and the investor's financial return for his investment. The prospective customer, before buying the wine, will have an eye on the bottle, and part of his/her decision to buy that wine will depend on how it looks. Furthermore, when the bottle is opened for tasting, it must meet the customer's expectations. In addition, the wine is kept in the winery cellar or warehouse for a certain period of time, and then at the store and the customer's home before it is opened and consumed. Care must be taken to ensure that the wine will age in the bottle without spoiling. All these factors must be taken into account during bottling. In this chapter we shall cover the important aspects of bottling, which are in principle the same at any scale, from amateur winemaking up to full-scale winery operations. The only difference is in the equipment (hand operated, semi-automatic, or fully automatic machinery), which is determined by the quantities to be bottled.

1. Bottles

Traditionally, wine bottles are made in three major shapes : the *Bordeaux* type, the *Burgundy* type, and the *Alsace* type (see figure O.1). The origin of these shapes is of course the bottle's

Fig. O.1 - Traditional wine bottle shapes.

name. Alsace type may also be called the "Hock" style. According to tradition, each varietal wine is held in a certain type of bottle. It would seem odd, for example, to find Cabernet Sauvignon in a Burgundy bottle or Chardonnay in Alsace or Bordeaux bottles.

The colors of the Bordeaux and Burgundy bottles are green for red wines, and white or greenish-white for white wines. The Alsace bottle, which is suitable only for white wines, comes in two colors : green in the Moselle style, and brown in the Rhine style.

It is important to mention that light may damage the wine, especially red wines. This is the reason why red wines are always bottled in deep green or brown (Italian style) bottles. Some white wines, which are intended to age for a long time, may also be bottled in green bottles.

The types of bottles, their color and the varietal wines which are bottled within, are summarized in Table O.1.

Type of Bottle	Varietal Wine	Color of Bottle
Bordeaux	Cabernet Sauvignon Merlot Zinfandel Sauvignon Blanc Semillon French Colombard Muscat	deep green deep green deep green white or green white or green white white
Burgundy	Pinot Noir Chardonnay Petite Sirah Gammay Chenin Blanc	deep green greenish white deep green deep green white
Alsace	White Riesling Gewurztraminer Chenin Blanc Muscat Sylvaner	brown brown brown or green brown or green green

Table 0.1 - Traditional relationship between bottle type and wine variety.

Varieties which are closely related to those cited in the table should be bottled in the same type of bottle as the one they are related to. For example, the Riesling family such as Emerald-Riesling (Riesling x Muscat) or Muller-Thurgau (Riesling x Sylvaner) are bottled in Alsace bottles. Practically all white German wines are bottled in Alsace bottles, as all German white varieties are Muscat related. The same goes for Cabernet-Franc and Merlot, which are associated with the Bordeaux region, and if released as varietals, should be bottled in Bordeaux bottles. The same argument holds for all Pinot related varieties (e.g. Pinot-Gris, Pinot-Blanc), which should be bottled in Burgundy bottles.

The common bottle size is, by worldwide agreement, 750 milliliters. There is also the half bottle (375 milliliter), the magnum (1.5 L), double magnum (3.0 L), and Imperial (6.0 L) sizes.

The neck of all bottle types and sizes (except the Imperial) has standard dimensions in order to fit the standard cork size. At the opening of the bottle, the internal diameter is 18 - 19 mm, and at 50 mm down into the bottle, it is 20 - 21 mm (see figure O.2). So the inside of the bottle's neck is tapered toward the top. This configuration of the neck holds the cork firmly in place.

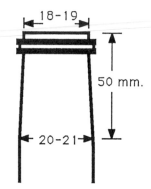

Fig. O.2 - Wine's bottle opening (shape and dimensions)

2. Corks

The cork is made from the bark of the *Quercus suber* oak tree. The tree grows in particular in Portugal, Spain, Algeria, Italy and other Mediterranean countries. Portugal is the biggest producer, with over half of the world production. Spain is the second producer with about 20% of the market. The bark of the tree is stripped about every 9 - 10 years, to produce the cork stopper. The first bark harvest which is of good quality for wine corks, can be collected 40 - 45 years after the planting of the trees. The annual world production is about 10 billion cork stoppers.

The production of the cork follows these steps: stripping the bark and leaving it in the sun and rain for about a year. Then the bark is boiled in water, which makes it softer, reducing its tannin, and eliminates all insect and microorganisms. Then the bark pieces are cut into strips and punched to cork stoppers. After washing, the corks are dried and get surface treatment for good insertion into the bottle. Over 80% of its volume is filled with air, and hence, it has very low density, high compressibility, elasticity, and some permeability to gases. Traditionally, and for good reasons, it is the only stopper used for quality wine all over the world, especially for long-aging wines. In principle, for fresh fruity wines, that are to be consumed young, there are cheaper alternatives, like plastic corks, crown caps or aluminium screw-caps. However in the case of high-quality wines, sold at medium to high prices, one can afford to use more expensive traditional corking. Pulling the cork from the bottle of wine is part of the ritual of wine drinking, it is as much a part of the style and appearance of the product, as the right bottle, the color and the label. The winery's logo is usually printed on the side of the cork.

For many years of bottle aging, no material is known to be better than cork. However, after a long time, the cork may deteriorate and fragment, causing the wine to leak out and allow air to enter. Very expensive wines (Bordeaux, Premiers-Cru category) which are aged by collectors for many years, must be opened and re-corked every 30 - 40 years.

There are different kinds of cork qualities. A good cork should be the right size, soft and elastic, smooth on its surface, without defects, and at about 6% moisture. Good quality cork has a specific gravity of about 0.13 - 0.16 gram/cm^3. As the specific gravity is higher, its quality is usually poorer.

The standard cork diameter is 23 - 24 mm, and when sitting inside the bottle neck (average internal diameter 20 mm), it is compressed to about 85% of its volume. The length range of corks is 40 - 48 mm for still wines, and over 50 mm for sparkling wines. For wines which are intended to age for a long time, the longer the cork, the better. Some corks consist of a few layers of cork pieces glued together or made of very tiny cork particles (conglomerate), held together by glue and paraffin coated. These corks are cheaper but not necessarily of lower quality than single-piece corks.

Corks may contain mold, yeast and cork dust; so it is advisable to rinse the corks in a 1,000 ppm solution of sulfur dioxide (0.1%) and 0.1% citric acid for about an hour.

However, when the cork is wet and has been squeezed by the corking machine, a brown solution of cork-tannin is extracted from the cork into the bottle. To prevent this from happening, after soaking the corks in the sulfur-dioxide solution, let them dry out for a day in clean dry containers before putting them into the corking machine. It is now possible to get washed and sterile corks in closed packages. In this case, the soaking can be avoided. Bad corks have an enormous impact on the quality of the wine and its appearance. It may leak or get moldy, or transfer to the wine a "corky taste". One of the substances that has been found to be responsible for such a corky taste, is 2-4-6 trichloroanisole. This compound needs chlorine for its synthesis, so it probably comes from the bleaching of the corks during their production. When buying corks, it is advisable to immerse a bunch of corks in 10% alcohol solution for couple of days, and to smell them carefully for hints of moldiness. If such a smell can be detected, do not buy that brand. Bad quality corks also tend to break down and disintegrate when the bottle is opened. The customers hate it. Try to avoid such corks in your production.

During corking, the air in the bottle is compressed, and if the corking machine does not operate under vacuum, the pressure remains in the bottle. If soon after the bottles are placed on their sides (a good practice for keeping wine bottles), the pressure may cause some bottles to leak. To avoid this, leave the bottles to stand after bottling for about a day or two to release the pressure before laying them down. It is much preferred to have the corking machine operate under vacuum when the cork is pushed into the bottle.

3. Capsules

Capsules are used to cover the cork surface and serve three functions: protecting the cork from cork borers, improving the bottle's appearance, and providing brand identity or authenticity. Traditionally, they have been made either of lead, or wax. Nowadays, they are made of tin foil, aluminum or plastic. The plastic capsules are made either with thick walls and are fitted to the bottle's neck by machine, or thin walls which are shrunk under heat to fit tightly around the neck. The winery's logo is usually printed on the capsule.

4. Bottling line

The bottling line assembly is shown in figure O.3. It starts with a receiving table for empty new bottles, which are taken out of their cases on this table.

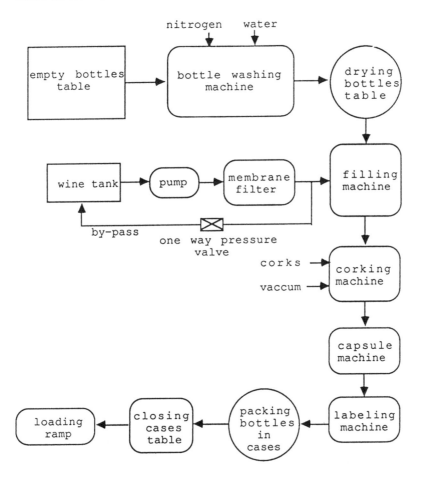

Fig. 0.3 -   Bottling line block diagram containing all necessary functions for wine bottling.

Although the bottles are new, they may contain some dust, therefore washing them is necessary. White bottles should be washed with filtered water, followed by a short jet of nitrogen. Colored bottles (green or brown) can be washed by nitrogen jet only, because any very fine dust particles are less visible in the colored bottles, and would not be seen in red wine.

Next to the washing machine should be a table with holes on its plate to place the bottles upside down for drying. It is very convenient for the workers in the line if this table rotates on its center. From that table the bottles are fed into the filling machine, which usually operates 6 - 12 filling stations, at an adjustable rate.

When a bottle in the filling machine is full, the filling is halted either by a floating device controlling the liquid level in the bottle, or by equalizing the filling machine reservoir with the liquid in the bottle. The filling machine is fed with wine which comes from the tank by a pump pressure through membrane filter of 0.45 micron size. It is recommended to place two membrane filters in series, the first one 0.65 micron and the next 0.45 micron, to assure a smooth flow of the wine without plugging the 0.45 micron filter. Normal working pressure on the membrane filters is between 0 - 15 p.s.i. When the pressure rises above 30 - 40 psi the filter should be replaced by a spare one, and at the end of the day should be cleaned by reverse flowing water (see chapter L.3). It is not recommended to use a pulse pump to filter the wine because of the pulsed pressure being exerted on the filter. Good filtration should be done by smooth and steady flow of the liquid. It is also recommended to flow the wine at the beginning of the day for about 10 minutes through the filter and through a by-pass line back to the tank, before starting the bottling. Between the filter and the filling machine reservoir (which is controlled by a floating valve), it is necessary to have a by-pass line back to the tank in case the filling machine reservoir is full. The one-way pressure valve on the by-pass will open the line when the pressure in the line rises above certain value (which happens when the filling reservoir is full).

The next stage after filling is corking. This is done by a machine with "jaws" that contract the cork before it is pushed into the bottle. The pressure needed is about 5 to 6 atmospheres. After the cork is pushed into the bottle's neck, it expands back to the neck walls and is held by the high friction constant of the cork surface and by the conical shape of the bottle's neck. The contraction should be done on the whole surface side of the cork with even pressure, otherwise the cork may be damaged and tiny chips of cork will be released into the bottle. The cork should be pushed into the bottle exactly to the level of the bottle opening. The headspace between the wine and the cork should be about 1 - 2 cm. Not less, because of temperature expansion, and preferably not more because of excess oxygen in the headspace.

During corking a pressure may develop in the bottle (because the cork is pushed as a piston) which can rise up to 20 - 25 p.s.i. It is preferred to have a corking machine which exerts vacuum in the bottle before pushing the cork in.

The labeling machine is the "bottle neck" of the line. Most of the problems in the bottling line are caused by this machine. It can apply three kinds of labels : the main label, the back label and the neck label. If one uses a neck label on his bottles, he should know that this label causes the most difficulties in labeling.

Last in the line is the packaging. The bottles should be packed in the cases neck down. If bottling is done manually, the speed is quite slow, whereas a typical automatic bottling line speed in a small winery is about 30 - 60 bottles per minute.

Some general remarks regarding the bottling line :

• If the wine to be bottled has some CO_2 in it (e.g. young white wine), it can be easily washed out by nitrogen. This is done by bubbling nitrogen through bubble-stone placed in the wine tank for 5 - 10 minutes.

• If there are any problems in the bottling line and the bottling has to be stopped, the first machine in the line - the filling machine, should be stopped, regardless of where the problem is.

• After each machine in the bottling line, it is recommended to leave some space for bottles to accumulate in case of a problem which may happen in the following machine.

• Wine from bottles which have been rejected from the bottling line, can be unified and returned to the tank after washing with nitrogen gas.

5. Sterilization

After passing through the filtration array, the wine is presumably sterile. This is meaningless, if the wine container namely the bottle and the cork, is not clean and sterile. Although directly out of production the bottles are sterile, by the time they are used for bottling they may acquire dust, mold and yeast cells. The dust is removed by a nitrogen stream and water. If sterilization is desired, it can be done either by rinsing the bottles with a 500 ppm sulfur-dioxide solution just before bottling, and letting water run out from the bottle for a couple of minutes, or by a stream of ozone followed with nitrogen to wash the ozone out of the bottle.

It is emphatically recommended to never bottle in used bottles. A specific cleaning machine for used bottles exists, but even with its use it is difficult to avoid a high percentage of dirty or unsterile bottles.

The bottling machine and the membrane filter must be sterilized every working day before and after the filling. This is done by passing through the filter and the filling machine hot water at 80°C for twenty minutes. When starting the bottling, the first few bottles may be partially mixed with water and should be rejected from the line.

If there is a need to touch the filling machine or the corking machine in order to fix a problem during bottling, after the problem has been fixed it is recommended to spray the area with 70% alcohol solution from a spraying bottle. The spraying can also be done from time to time without any interruption, just to assure that those parts of the machine which come into contact with the wine, are sterile.

6. Labeling

The label is the front yard of the wine. It has an enormous impact on the potential customer. Besides being aesthetic, the label should contain sufficient information about the wine. This includes the name of the winery, the type of wine (varietal or generic), and whether the wine is a table wine or a dessert wine. If it is a table wine with some noticable residual sugar, this should be indicated. Also, by law in most countries, it should be clear from the label what the alcohol content is (v/v), as well as the wine volume. In the United States, as of January 1987, the label must note that the wine contains sulfur-dioxide if there is more than 10 ppm of total SO_2.

More important information is the variety, the vintage and the appellation of the grapes. In each country there are specific rules concerning these issues. California's rules are given in chapter L.4 on blending. The label has to be approved by the authorities in most countries before the wine carrying that label can be released.

Some wineries put a back label on the bottle bearing additional information about the wine. This information is not formal, and usually includes some details on the production process, the variety style, or some analytical data on the grapes and/or the wine.

If the wine has to be aged in the winery cellar for a couple of months or longer, it is recommended not to label the bottles right after bottling, and let them age unlabeled. Just before releasing the wine the bottles are labeled. By doing so, one can avoid the problem of damaged labels caused by the long storage.

Finally, a note on the glue for labeling white or blush wines, which are chilled before use. If the glue is water-based, the label may fall off during or after chilling; therefore, it is advised to use non-water based glue.

7. Bottling Storage and Aging

White Wines - can be divided into two groups : fresh and fruity wines which should be drunk young, and those which have good potential for aging and can be aged and improved for a couple of years before drinking. The first group of white wines includes varieties such as Chenin Blanc, Gewurztraminer, Riesling, Muscat, French Colombard and whites made of red grapes (Blanc de Noir). These wines, after recovery from bottling for a couple of weeks, are ready to be released to the market. The second group of white wines, which include Sauvignon Blanc, Chardonnay and Semillon, improve by aging in the bottle for one or two years before being released. The question of how long to age these wines before release is an economical one.

Red Wines - are generally bottled about one and a half to three years after harvest, depending on how long it has aged in the barrel. In general, after bottling, the wine has to age for one to three years in order to develop bottle bouquet, which is a significant parameter in red wine quality. If the winemaker finds a premium red wine lot, it may be worth aging in bottles for as many years as needed, and then sold for a price which justifies the long period of storage. Otherwise, the wine can be released a few months after bottling, according to the market demand.

Light red wines made in a style to be consumed young (Beaujolais Nouveau style) are released as soon as the wine has finished clarifying and stabalization.

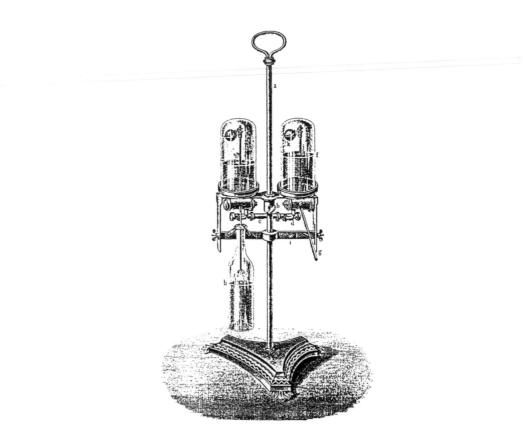

P. QUALITY CONTROL

The definition of quality control which is presented in this chapter has nothing to do with the wine quality itself, but a set of measures which are taken to assure that the product meets certain specifications. These specifications (chemical composition, microbiological content, chemical stability, packaging standards and others) are structured to fall within certain levels of measurable tolerance.

We also refer to quality control only with regard to the final product and not to the production process. The latter is a very basic part of wine making and has been described in depth in the previous chapters of the book. The quality control of the final product is a kind of screening stage where only those items which meet the pre-set standards are allowed to pass through. These standards are in part enforced by legal regulations and in part by the winery itself.

In large industrial plants (including big wineries), the production department and the quality control department are sometimes caught up in a conflict of interest. To avoid this, in a well-managed organization, quality control is separated from production, and is conducted with a direct reporting channel to top management. In small wineries where the full responsibility of the production lies with the winemaker, this is impractical because of the small size of the plant, therefore it is his responsibility to set standards and carry out the quality control according to his best judgement. His professional reputation as a winemaker is re-evaluated with every new vintage released to the market.

A major goal in food industry quality control is to maintain consistency of the products. When a customer buys a food product, generally he or she knows what they are buying from previous experience. The customer can judge very quickly any change in the taste, smell, or aroma of his favorite brand. In the wine industry, consistency is more difficult to achieve. In very big wineries, consistency is possible because of the enormous quantities of wines which can be blended to meet, as close as possible, the consistency required. In small to medium wineries, there is no way to do so and, basically, no need for consistency. The goal of the winemaker in the small or medium size winery is to do his or her best to process the vintage grapes into the highest quality wine possible. Specifically, quality control is a series of tests and inspections incorporating certain procedures which deal with different aspects of the wine. Some details on the procedures mentioned in this chapter (the analytical ones) are found in chapter R.

1. Legal Limits

The legal limits in which we shall deal with here are only those which are important at the final product stage (not during processing). The distinction is important because there are regulations which are imposed during the processing, e.g. sugar addition to the must before or after fermentation, fining regulations, chemical additive regulations and more. These regulations, which may be legal in one country might be illegal in others.

The most important regulations of the final product are the alcohol content, the sulfur-dioxide, and the volatile acids concentration. The definitions and limits are specific for each country. Here we refer (as an example) to the United State's regulations, which are controlled by the Bureau of Alcohol, Tobacco and Firearms (BATF):

i. The legal limits of alcohol in table wines are 7 to 14% of alcohol (v/v). The 7% is the minimum needed to be considered as wine, and the 14% indicates the maximum value to be considered as table wine for tax classification. There is no flexibility in the measure of the upper limit. If the wine is checked by the authorities (BATF), and the alcohol measure is found to be in excess of the exact 14% limit, it automatically enters the next category of alcohol taxation (fortified wine 14% - 21% alcohol), which is about four times higher than table wines. It is not mandatory to present on the label the exact alcohol content (although most wineries do). Instead of alcohol content, it may state that it is a "table wine" and it is understood that this wine contains alcohol which lies in the range between 7% to 14%. If the analysis of the alcohol content is very close to 14% according to the measurement, the winemaker is advised to be sure that it is really below 14%, and if in doubt, blend it to a safe distance from that limit if the wine is to be sold under the table wine taxation category. In some varieties the sugar content at harvest is high (above 24 B°), and as a result, the alcohol concentration may be above 14%. It is accepted by the market that these varieties (e.g. Chardonnay, Zinfandel) may contain such high alcohol concentration, and in some wineries this is part of their style. Although these are table wines by definition, they are classified by tax regulations as fortified wine.

ii. Sulfur-dioxide is one of the oldest chemicals used in wine production. Its importance in the process is discussed in great detail in chapter D, which is dedicated to this substance. Its use is not unique to wine production but it is in use in many other food products. In wine, the total sulfur-dioxide limit in the United States and in most countries, is 350 mg/L, which is very much higher, in general, than the concentrations found in practical use. In dry fruits, for example, the limit of sulfur-dioxide is 2000 mg/kg. In recent years, questions have been raised regarding the possible health risks by using sulfurdioxide. There is no conclusive evidence of any risk by using sulfites at the legal limits, excluding very few allergic people who are sensitive to this particular substance. But even though the general use of sulfur-dioxide in the wine industry is very much below the legal limits, there are, and have been, attempts to minimize it even more or to replace it with some other substance. As a consequence, in recent years there has been a significant trend to reduce the use of SO_2 in the United States wine industry. With regard to the analysis of SO_2 concentration for the purpose of legal limits, only total SO_2 is of concern. As for the protection aspect, only free SO_2 is important, and it is general practice to bottle the wine with 20 to 30 ppm of free SO_2. In the United States any wine containing more than 10 ppm of SO_2, must have a note on its label saying "contains sulfites".

iii. Volatile Acidity (mainly acetic acid), may be found in wine for one of several reasons : as a by-product of the fermentation generally in the range of 300 to 400 mg/L, and as a major product of Acetobacter infection, with very high concentrations. It may also be produced from citric acid during ML fermentation. The other volatile acids (formic, propionic, butyric), are present in very low concentrations. The analysis of volatile acidity contains all these acids, and is expressed as acetic acid.

The legal limits of volatile acidity in some countries are given in Table P.1 :

	U.S.	California	France	W. Germany
White Wine	1.2	1.1	1.1	1.2
Red Wine	1.4	1.2	1.1	1.6

Table P.1 - Maximum legal limits of volatile acidity (in gram/L) in different countries.

Since the analytical method of determining volatile acidity is by steam distillation, some other volatile acidic substances are also distilled, such as free SO_2, CO_2 and sorbic acid. These substances have to be subtracted from the total distillate analysis, in order to have the right concentration of the volatile acidity.

2. Instability

This term includes two major sections :

a. chemical instability factors responsible for changes occuring in the bottle such as cloudiness, precipitation and color changes, which appear under certain storage conditions.

The chemical instability factors are : proteins instability, tartrate instability, metallic haze (iron and copper), and polyphenols (color changes and precipitation).

b. microbiological instability, which can cause major changes in the wine's composition due to microorganism factors.

These factors consist of certain kinds of microorganisms (yeast and bacteria) which under certain conditions (pH, temperature, nutrients, oxygen, lack of protecting agents), may begin to multiply and change the chemical stability, causing off-flavors and changes in the wine's character.

a. Chemical Instability

One of the most important goals of quality control is to prevent instability hazards, and if found, to identify the factor and treat it before bottling. The general technique to test for chemical instability is to perform a set of temperature changes according to this scheme : A couple of corked bottles are heated at 60°C for two days, then cooled to room temperature for a day, and then for two days at -5°C, and back to room temperature. After a day at the final room temperature, if cloudiness can be seen by direct or indirect light, there is an instability problem present in that wine. The cause may be one of the following : protein coagulation, iron or copper casse, or tartrate crystallization (possibly calcium tartrate). To identify the factor, a series of selective tests has to be done. The most abundant problem is protein instability in white wines. It can be checked by two methods. First, by heating the wine (a new clear sample) at 80°C for 6 hours and cooling back to room temperature. If no haze exists (in a few days), the reason is not protein. If the wine is cloudy, protein instability is most likely the case. The second test (following the first one) is to add to a 10 milliliter wine sample, 1 milliliter of trichloroacetic acid and heat it in a boiling water bath for two minutes. Cool to room temperature. A precipitation or haze indicates protein instability.

Protein instability can be treated and cured with a fining agent (see chapter M).

Iron casse is a result of oxidation of ferrous ion (F^{+2}) to ferric ion (F^{+3}), which precipitates as ferric phosphate at low pH 3.0 - 3.5 and cold temperatures. Iron casse may happen when its concentration is above 5 to 10 mg/L. If, by heating a cloudy bottle of the tested wine back to 60°C for a couple of hours and the haze disappears, this is probably the case. To eliminate the possibility that it is not tartrate instability, add to that cleared bottle 1 milliliter of citric acid, which interacts with the iron ion to form a soluble stable complex. Cool to -5°C for a day and return to room temperature. If there is no haze, the reason is an iron casse.

Copper casse can be developed in the absence of oxygen and, with sulfur-dioxide presence precipitates as copper sulfide. By oxidation, the sulfide may transform to sulphate which is soluble.

The copper casse may happen when the copper concentration is above 0.5 mg/L. The test for copper casse is very simple. Aerate the wine from the hazed bottle, if the haze disappears, it is most probably a copper casse.

To treat the wine in either the iron or the copper casse, one has to reduce their concentration below the corresponding figures above. This can be done with commercial substances, aimed to remove both metals at the same time (e.g. substance called Cufex, which at 100 mg/L can reduce the copper or iron concentration by 1 mg/L; manufacturers instructions should be read carefully before use). An alternative for iron casse prevention, most frequently recommended, is addition of citric acid at no more than 300 mg/L to the wine. Laboratory tests to determine the minimum amount needed should be done.

If, the wine becomes cloudy, by being cooled to -5°C and if the tests for iron and copper are negative, then the reason is most probably tartarate instability (in spite of the cold stabilization which was done early during the wine processing). The reason is probably calcium-tartarate instability (see chapter L.2 for details). After verification, the best treatment is to cold stabilize the wine once more, and to facilitate the crystallization, seed the wine with calcium tartarate seeds and filter cold.

Polyphenols instability in red wines is a natural process of oxidation and aging, and no measures have to be taken because it will happen in any case after years of storage. The polymerized polyphenols form dark-red sediment which is not considered a defect in the wine. When it happens in old red wine, the wine should be decanted just before serving.

Color changes are mainly due to oxidation of the phenolic compounds in the wine. In white wines, the color turns yellow, and in red wines to brown. In "blush wines" or rosé, the purple color changes to brown-purple. After long bottle aging, these changes will happen in any case. The objective is that these color changes will not happen in a short time after bottling (in white wines and "blush" for a couple of months up to one year, and red wines in 3-5 years).

There is no precise method to forecast these color changes. The best that can be done is to accelerate the process by placing some corked bottles at high temperatures (40 - 50°C) for a couple of weeks, and compare the color to the untreated wine. If the color changes look serious, the only way to delay it is by reducing the oxidation potential in the wine. This can be done by adding some reducing agent such as SO_2 or ascorbic acid before bottling.

For more details on chemical disorders, see chapter Q.2.

b. Microbiological Instability

Microbiological instability can be caused by certain kinds of yeast species and bacteria. For details see chapter Q.1 on wine spoilage.

3. Bottling Quality Control

The bottling quality control consists of a set of tests which are conducted during the bottling phase on bottles selected at random, and also by individual inspection of the bottles coming out of the bottling line.

A very basic principle to remember is not to bottle a wine unless full analysis of these parameters have been made : alcohol, total acidity, pH, volatile acidity, residual sugar, sulfur dioxide, dry extract, and the stability tests.

The sterility of the wine has to be checked after it leaves the bottling line by taking samples at least once at the beginning of the bottling day and once at the end. The test can be done very quickly by using a microscope (see chapter R) to detect yeast cells, bacteria, lees particles, or filter segments. This test shows the filter's condition and will detect whether there is any breakage in the membrane filter (which occurs occasionally).

A second test, which requires more time to get results, is to put aside two to three bottles from the line, empty half of each, plug a piece of clean cotton into its neck, and leave for four days in the laboratory at room temperature. Check these samples for volatile acid, total acidity, and by tasting. This test gives the winemaker qualitative information about the sterilization of the wine. If, after four days, there is no acetobacter infection in the bottle, it shows that the wine was sterile filtered. It is also recommended to take a sample at each bottling day and run a quick aging test at 50°C for a week. Following this, open and taste the sample, and, if there is something unusual (besides aging effects), proceed with full analysis to determine what is wrong.

On the bottling line, after bottling, corking and capsuling, but before labeling, the bottles should be inspected with direct light for particles within the bottle. Any suspected bottle should be rejected from the line. At the same inspection, the level of the wine in the bottle should be watched. Any noticeable deviation from the set-up level should also be selected out. The same for defective cork or capsule. The labeling operation should be carried out with individual inspection of each bottle. Quite frequently the labels are not set in the right place on the bottle, are torn-down, or dirty, two labels are stuck together, or a variety of other defects. These bottles should be selected out and re-labeled.

The aesthetic appearance of the bottle and label is most essential. Before testing the wine, the potential customer judges the bottle from the outside. Even if he or she became intent on buying your wine, but saw a defective label, defective capsule, unclean bottle, or any other unaesthetic element, most probably the prospective buyer will change their mind. This bottle, in the end will return to the winery. The rule is simple - for a quality product, the packaging is as important as the product itself.

4. Miscellaneous

i. Chemical Management in the Vineyard - In general, unless the winery owns the vineyard, the winemaker has no information as to when and what kinds of chemicals have been used in the vineyard which may affect the wine processing (chapter B.1). He is however responsible for any health hazards which may be caused from chemical residues that had been left on the grapes and were carried into the wine. It is, therefore, most important that the winemaker, when buying grapes from the grower, demand to have the full records of the chemical management (fungicide, herbicide, insecticide, anti-virus, sulfur and any other), that have been used in the vineyard during the current year right up to the date of harvest. The list of the chemicals in use, their characteristics in terms of hazards, stability and related vineyard management practice should be familiar to him, in order to be fully aware of any possible problems.

ii. Analytical Records - All the analysis taken during any lot processing, and all treatments and chemical additives, from harvest to bottling - should be recorded and followed on each lot of wine. In case of any problems which may arise in the winery or after the wine has been shipped out - there are references to check in order to determine what went wrong. For an example of such record forms, see Appendix D.

iii. "Organic" or "Natural" Wine - As part of a certain trend in attitudes toward food growing and processing - namely, "natural" or "organic", as they are called, there is some demand and supply of wine which is "naturally" processed. This includes vineyard management without chemicals, even without fertilizers, no use of sulfur-dioxide at any stage of processing or of any other chemical additive, no use of any fining agents, and no use of filtering. Also, no addition of any types of yeast or ML culture.

In principle, it is possible to produce wine without using any of the modern technology methods. The price paid is likely to be the following :

- A good chance of the wine spoiling, either during processing or later, in the bottle.

- A high degree of instability, both chemical and micro-biological.
- Imbalance in terms of taste and quality.
- Probable over-oxidization.
- A very unpredictable wine in terms of what will come out when the bottle is opened.

In regard to this "natural technology", my personal response could not be better said than by quoting the following section from the early nineteenth century book, "History and Description of Modern Wines," by Cyrus Redding (1833) :

> "The wines of moderns are, there is no doubt, much more perfect than those of the ancients, as far as can be discovered by any thing authentic, which has reached the present time".

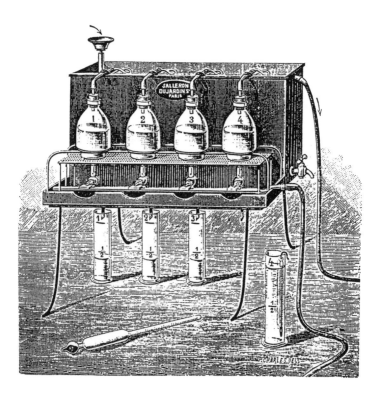

Q. WINE SPOILAGE

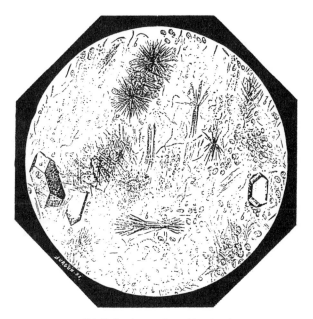

Maladie des *vins tournés, montés,* qui ont la *pousse.*

Almost everyone has opened a promising bottle of wine and found it to be unpleasant, repelling or even undrinkable. In chapter P.2 turbidity and sedimental instability are discussed. In this chapter we touch on the nose and flavor problems which may spoil wines. The causes might be a chemical spoilage, or in most cases a microbiological agent. The discussion will be divided into these two factors:

1. Microbial Spoilage

Although no known pathogenic microorganism can survive in wine media, many others can grow and flourish very well in such low pH and high alcohol environment. Here we shall mention some of the most common agents which cause wine spoilage.

A distinction has to be made between two terms which are used in regard to microbiological activity in wines, *fermentative* and *oxidative.* The *fermentative* term refers to anaerobic decomposition of carbohydrate derivatives by the microorganisms (yeast or bacteria), and the *oxidative* term refers to aerobic oxidation of these substrates. The microorganisms which are active in wine spoilage belong to both groups.

The factors which affect the growth of microorganisms in wine are:

a. pH

In the wine pH range 3.0 - 4.0, the lower the pH, the less likely the growth conditions of most spoilage microorganisms (except acetobacter). Most species of lactic acid bacteria, for example, are unable to grow below pH 3.2.

b. Alcohol

The alcohol is an inhibitor at different concentrations to most microorganisms. Some yeast strains can tolerate up to 18% alcohol, where other wine microorganisms are sensitive to low concentrations of about 8% .

c. Temperature

The preferred temperature range for most spoilage micro-organisms is between 20 - 35°C, but they can also grow slowly at low temperatures down to 10°C and below.

d. Sulfur dioxide

Sulphur dioxide is the main added chemical used as an anti-septic agent in must and wine. The only part of the sulfur-dioxide which has protecting power is the free molecular SO_2. The different microorganisms have various resistance to its presence and concentration. For full details on sulfur-dioxide, see chapter D.

e. Residual Sugar

The existence of fermentable sugar (above 0.1%) in the wine, facilitates the growth and the risk of spoilage. High sugar concentrations 10% - 20%, in combination with alcohol make the conditions for microorganism's growth unfavorable.

f. Growth Factors (nutrient)

This includes different ingredients which are necessary for the various species of microorganizms to grow and to be active, such as amino-acids, vitamins and minerals.

g. Air

Essential for the growth of aerobic microorganisms.

Yeast

• Yeast is regarded as spoilage if it is active after the desired sugar/alcohol fermentation of the must (in still wines). The spoilage can happen in the tank during the wine processing or it can happen in the bottle. In both cases it causes undesired changes.

The basic factor which is necessary in order to enable yeast growth is fermentable sugar at any concentration above 0.1- 0.2% (not including the unfermentable pentose sugars which are left over in the dry wine at concentration of about 0.2 - 0.3%). This sugar may be left over after fermentation or may be added to sweeten a dry white wine. In both cases, when the wine is in the tank waiting to finish its processing (clarification, fining, aging), if no measures are taken to prevent it, the wine will referment. The most effective measure at such a stage is to keep the wine at a low temperature, in the range of 4 - 7°C. The low nutrient level, the alcohol and the low temperature will inhibit the yeast growth. Sulfur-dioxide has practically no effect at its regular concentration on their growth. As long as the wine is kept at such a low temperature, there is no reason to expect any problems. However, if for some reason, the temperature rises, the refermentation will be inevitable. The first sign would be tiny bubbles of CO_2 in the wine, or a "spritz" feeling on the tongue, when tasting a sample of that wine. A microscopic test of the wine sample will show the budding yeast.

If no immediate measures are taken, the entire residual sugar will ferment, causing a serious problem. The best way to stop it is to filter the wine through a medium-fine filter 40 - 70 pad to reduce the yeast population, and to cool the filtrate to below 7°C.

A more serious problem is the bottle fermentation of semi-dry or off-dry wines. First there is a physical risk of bottle explosion under the CO_2 pressure, which may rise up to 6 - 7 bars (where the fermentation will stop under the high pressure). Also the slightest signs of fermentation in the bottle, usually destroys the customer's good feeling about that wine. For more details on how to prevent the bottle fermentation, see chapter K. In red wines, this problem practically does not exist, as red wines for all practical organoleptic purposes, do not contain fermentable residual sugar.

• Another yeast group which infects wine is called surface-film yeast. This yeast belongs to many yeast genera, including the Flor Sherry yeast (*Saccharomyces fermentati*), which are used for the production of sherry.

The common characteristics of this yeast are :

a. they are aerobic and need oxygen for their growth ;

b. they produce colonies on the wine surface ;

c. they oxydize alcohol to produce acetaldehyde which resembles sherry wines. If the infection is not treated, this nose will grow up to an unpleasant off-flavor bouquet, which is undesired in table wines.

The surface-film yeast grows only when the wine surface is in contact with air, and when the SO_2 level is low. Free SO_2 at concentration of 20 - 25 ppm will prevent its growth. If such infection has been discovered, the treatment should be SO_2 addition, and prevention of air head space above the wine. This can be done either by filling the tank to its top or by removing the air and keeping nitrogen headspace.

• One more remark on a special strain of yeast, *Saccaharomyces bailii*, which may spoil wine even with the presence of high concentration of inhibitors such as 200 ppm of SO_2 and 200 ppm of sorbic acid. This yeast can also stand a very high concentration of sugar in grape concentrate, up to 70%. The polyphenols in red wines and alcohol above 12% inhibits their growth. This osmophilic yeast strain which exists amongst natural wine yeast have large ellipsoidal cells in the range of 5 - 10 micron. Due to their survival strength, it is difficult to fight these yeast (fortunately quite rare), and the best protection is sterile filtration.

ML Bacteria

As mentioned in chapter J.4, there are three major genus living in wine medium (low pH, high alcohol), and the *Leuconostoc oenos* are the desired ones for ML-fermentation. The species belonging to the two other genus, *Pediococcus* and *Lactobacillus* are undesirable and cause wine spoilage whenever they manage to grow in the wine.

• The *Pediococcus* (which are round shaped bacteria, bigger in size than the *Leuconostoc*, and appear usually in pairs or clusters under the microscope), are responsible for a very unpleasant "wet socks" smell. They live on malic acid, even if its concentration is very low.

• The *Lactobacillus* (long rod shaped bacteria which usually appear in long chains) live on lactic acid but can consume glucose and citric acid to produce acetic acid. It can also transform sorbic acid if present in the wine, to a "geranium like" unpleasant smell (see chapter K).

• Even the *Leuconostoc* bacteria (which under the microscope appears as small oval cells attached in pairs at their early stage of development and later on as long chains), can cause problems if they ferment the wine when not desired, especially in the bottle. Red wine which went through ML-fermentation, but did not finish it completely, may do so in the bottle if the filtration was not sterile. The results are not as bad as the other genus, but the wine will then have CO_2 bubbles in it, together with some sediment. In fresh white wine, it will destroy its fruitiness and give it a buttery flavor.

All the measures that will discourage the growth of the ML bacteria, are detailed in chapter J.4. If the infection happens when undesired, it may be too late. The best way is to prevent it. In the bottle, it is best to filter the wine (0.45 micron), together with enough SO_2 addition prior to bottling.

Brettanomyces Yeast

This yeast, which belongs to the Brettanomyces genus is very common amongst the wild yeast existing in the winery. Its microscopic identification is difficult since it appears in many shapes, from oval to long sausage forms. It also exists in a sporulating and unsporulating forms. This yeast was first found in the brewery industry, causing a "secondary fermentation" which turbids the beer, giving it what is described as a maukish flavor. In wine it grows mainly in red wines, and rarely in whites. The contamination is carried through the winery equipment (tanks, barrels, filters, etc.), but which wine is susceptible to the infection is not clearly known. The yeast can be dormant for a very long time (even years), and then start its activity in the cellar or in the bottle. The result of the Brettanomyces activity in wine is a special smell which at the beginning in very small concentrations may be considered as "interesting" or "complex", but later it is described as a "horse blanket" or "wet dog" smell.

The yeast is sensitive to free SO_2 if its concentration is in the 20 - 30 ppm range. In red wines, where the pH is usually 3.4-3.6, more SO_2 is needed to maintain the 0.8 ppm of molecular SO_2 which is effective as antimicrobiological inhibitor (see figure D.1). To avoid its activity and to prevent its wine spoilage risk in the cellar it is necessary to maintain enough SO_2 level or if not, follow the wine very carefully for any sign of its presence (by smelling). In bottling, the sterile filtering (0.45 micron) will assure its elimination from the wine.

Acetobacter

There are many different species of the acetobacter genus. Two of them are mentioned here, *Acetobacter aceti,* and *Acetobacter oxydans.* The first species are found mostly on grapes, attacking the rotten and infected berries (their count in botrytised must might come up to 10^6 cells/ml). In healthy grapes their population is around 10^3 cells/ml. During fermentation their numbers decrease substantially, but at each racking it is increased. Their optimum temperature of activity is 30 - 35°C. The second species which tolerate alcohol very well, is found mainly in wine with optimum temperature activity of 20°C. The *Acetobacter-aceti* can oxidize alcohol all the way up to carbon-dioxide and water, where the *Acetobacter-oxydans* are able to go up to acetic acid only. Both species can grow on glucose, glycerol and high alcohols, but they prefer ethyl alcohol.

The acetobacter are aerobic and need air for their growth. If the wine surface is exposed to air, the bacteria which is always present in the wine, will start to grow near the surface by consuming alcohol and producing acetic acid and ethyl acetate. Acetic acid itself adds some sharpness to the wine which is felt mainly in the throat. In small concentrations, it does not necessary spoil the wine. In botrytised late harvest wine the concentration of volatile acidity (from the grapes) is quite high, giving the wine the pungency feeling. The accompanying compound to acetic acid in acetobacter activity - ethyl acetate is responsible for the unpleasant repelling smell of that infection. So, the volatile acidity measurement is just a sign for potential acetobacter infection and how much it has proceeded. In the normal process of winemaking the volatile acidity is produced by the yeast and ML-bacteria in the range of 300 - 400 mg/L. Above this range the alarm bells should ring.

Acetobacter tolerate alcohol (their main substrate) up to concentration of about 15%. At higher concentrations they cannot grow, and this fact is used in Sherry production to maintain the aerobic process of the Flor Sherry yeast without the interference of acetobacter spoilage. The wine is fortified up to minimum 15% alcohol before the Sherry production.

SO_2 in the regular concentration does not protect the wine against acetobacter, and due to the fact that they always exist in the wine and in the cellar, the only way to prevent their growth is to minimize air contact on the surface, and to keep the wine at a cool temperature of about 10°C.

2. Chemical Disorders

In this section we refer only to the existence of extra concentration of chemicals (as a result of excessive addition or careless treatment of the wine). The 'natural' chemical instability is described in chapter P.2 .

● **Sulfite** - Sulfur-dioxide is added to the wine during its processing from crushing up to bottling. These additions are cumulative and at the end, the wine in the bottle will contain a certain quantity of bound SO_2 (which is difficult to recognize by taste), and a certain amount of free SO_2 which is described as a burning sulfur or pungent smell. The average threshold of free SO_2 is about 40 - 50 ppm. In healthy grapes, after a certain amount of SO_2 is added, most of the extra additions remain for quite a long time as free SO_2. Hence, care should be taken to not have during bottling more than 25 - 35 ppm of SO_2, otherwise it may be recognized in the wine.

In white wines the total SO_2 under normal conditions is usually about 70 - 150 ppm. In infected and rotten grapes, the total SO_2 can be more than these values due to the high bounding compounds (mainly aldehydes) in the must and wine which hold more sulfur-dioxide in the bound form.

In red wines, the pigments can keep part of the SO_2 in a reversible equilibrium which makes the red wine more susceptible to higher concentrations of total SO_2 before it becomes a chemical spoilage problem.

Excessive free SO_2 rarely happens in modern winemaking, as the SO_2 regime is strictly controlled. However, if it does happen, the way to reduce the extra free SO_2 from the wine, is to strip it by aeration or by oxidation. The stripping can be done either by aeration racking (from the bottom of the tank to the top of another one), or by bubbling nitrogen through the wine for a couple of minutes (from the bottom of the tank). The aeration method is not efficient enough in the wine with a high pH. A good stripping is done at very low pH of about 0-1, as it is done in the analysis of free SO_2 in red wine by the aeration-oxidation method (see chapter R.4).

The other option is more efficient and is based on the oxidation of the SO_2 with hydrogen-peroxide:

$$SO_2 + H_2O_2 \text{--------}> SO_4^= + 2H^+$$

Use 1% and follow this guideline : To reduce the free SO_2 in the wine by 10 ppm, add 50 ml of 1% solution hydrogen-peroxide to 1HL of wine. An excess addition of peroxide may oxidize other components in the wine, reducing its quality. Therefore, the addition should be made only after a laboratory test has been performed to find out the exact volume of peroxide needed to reduce the SO_2 to the desired level without affecting the wine quality. The addition should be done in small steps, 10 ppm at a time.

• **Carbon Dioxide** - In young white or rosé wines which were kept cold in order to avoid fermentation up to bottling, it is possible that during bottling the wines contain enough CO_2 left over from the fermentation, to produce a "spritz" feeling on the tongue. In most cases this is not desired and the best way to eliminate this CO_2 from the wine is by nitrogen bubbling through the wine for a couple of minutes.

• **Sulfide** - If the formation of hydrogen sulfide was not treated during fermentation, or after the first racking, it will stay in the wine and may later transform into mercaptans and disulfides (see chapter J.1). The smell is described as "sewage", "rotten eggs", or "garlic". If disulfide has not yet formed, an easy test can verify the sulfide problem. Addition of one-two drops of 0.5% of copper-sulfate solution to a glass of suspected wine should eliminate that smell immediately after mixing and comparison to untreated glass will help. If such a problem has been discovered, and not treated during the fermentation period, it can be partially cured by addition of copper sulfate ($CuSO_4.5H_2O$) to the wine in the range of 0.05 - 0.5 ppm of copper. The copper reacts with the sulfide and mercaptan to produce insoluble copper silfide. If disulfides (R-S-S-R) are present in the wine, the copper will not react with it, and pre-action with ascorbic acid (10 to 40 ppm) can break the di-sulfide to mercaptan (R-SH), which can then be treated with copper. Before treating the wine, a laboratory test has to be performed to find out the right cause of the unpleasant odor, and the minimum value of copper addition which will cure the smell, keeping in mind that excess copper in the wine might cause copper haze instability (see chapter P.2).

The test is carried out with three glasses of 100 ml of suspected wine.

To the first glass add 5 drops (0.2 ml) of the copper-sulphate solution and mix well.

To the second glass add 5 drops of ascorbic acid solution (10 gm/L in distilled water), mix well, wait a couple of minutes and add 5 drops of the copper sulphate solution. Again mix well. The third glass serves as a reference. Swirl each glass (starting with the reference), and look for a change in the unpleasant odor. The following table will help in coming to conclusion:

Case	Glass 1 Copper	Glass 2 Ascorbic Acid/ Copper	Conclusion
1	No change in smell	No change in smell	Not a sulfide problem
2	No change in smell	Reduction or elimination of smell	Disulfide
3	Reduction of smell	Elimination of smell	H_2S, mercaptan and disulfide
4	Elimination of smell	Elimination of smell	H_2S and/or mercaptan

The cure treatment follows according to the conclusions. If disulfide is not present (case 4), the copper addition will help. If disulfide is also present (case 2 or 3), the combined treatment of ascorbic acid/copper has to take place.

The copper stock solution can be made by dissolving 4.1 gram/L of copper-sulfate in hot water. For the laboratory test, 10 ml of that solution should be diluted to 100 ml with distilled water in a 100 ml volumetric flask. To an array of 100 ml of wine samples add 0.05 ml, 0.1 ml, 0.2 ml,.....0.5 ml of the diluted copper solution, which represents 0.05, 0.1, 0.2, ... 0.5 ppm of copper addition, respectively. The result should be checked by smelling. When the right concentration of copper needed to cure the problem is found, one can use the copper stock solution (4.1 gram/L) according to the ratio :

> 0.1 ppm of copper = 10 ml of copper solution/ HL of wine

For example, if you found in the lab test that 0.2 ml of copper addition is necessary to cure the smell, then 0.2 ppm of copper is needed, so according to the above ratio, 20 ml of copper stock solution has to be added to 1HL of wine. For the combined treatment, usually 25 ppm of ascorbic acid is needed prior to the copper-sulphate addition, which is 2, 5 gm/HL of wine. The ascorbic acid should be dissolved in small amounts of water, then be added and well mixed in the wine. Wait a day before the copper addition.

● **Oxygen** - Too much oxygen in white wines will show up by a yellowish color, lack of fruitiness, flat aroma, bitterness and even a caramel flavor. In rosé wines or light reds, the color will be a pinkbrown or red-brown and usually have a flat aroma. Red wine pigments are oxidized and polymerized to lose their red color. They would age in the bottle quite rapidly, depending on the amount of tannins in the wine. Over-oxidation may also cause aldehyde formation. In order to prevent these results, minimum aeration should be practiced during wine processing, especially in white and "blush" wines. Red wines are less susceptible to oxygen, but not to intense and frequent oxidations.

In white wines, sulfur-dioxide and ascorbic acid additions may inhibit and delay the oxidation. Sulfur-dioxide has been discussed in details in chapter D. Ascorbic acid (vitamin C) can be added to the must during crushing, in quantities of 50 - 200 mg/L. Ascorbic acid, which is effective against non-enzymatic oxidation, alone or together with SO_2 is a very efficient antioxidant agent.

● **Over-processing** - This term includes operations in the course of the winemaking, which has been done carelessly or without paying attention to the impact on the wine quality. Such operations are filtering, fining, racking and oak aging.

Over-filtering (with D.E. or pad filtering) again and again, gives the wine a "papery" or "earthy" taste. Red wines should not be filtered at all, except before bottling with a sterile filter.

The best clarification is done during aging in the barrel by natural precipitation. White wines should be filtered once or twice after cold stabilization, before bottling filtration.

Fining with any agent (such as bentonite, gelatine, carbon, etc.) should be done with the minimum quantity required to fulfill its function, otherwise it will "strip" the wine of its aroma and flavor.

Over-racking will introduce too much oxygen to the wine, causing over oxidation.

Over-barrel-aging may cause the wine to taste and smell over-oaky which sometimes dominates the natural wine aroma and flavor. The wine is then unbalanced and its quality is definitely reduced. Tasting the wine during barrel aging is necessary to prevent such results. Also careful selection of the barrels to match the wine at hand is recommended, depending on the barrel's condition and age (for details see chapter N).

R. BASIC ANALYSIS OF MUST AND WINE

"The wine should have an absolute unity, or taste as one whole."

—On this point, the power of tasting is more effective than that of chemical testing. The latter can, within certain limits, determine the proportions of the constituents of a wine, but not ascertain whether they were derived from one source or from several."

* *From "Chemical Testing of Wines and Spirits" by John J. Griffin, F.C.S. (1872).*

In this chapter, we present the basic analytical practice of must and wine which the winemaker can perform in his own small laboratory. For a more detailed analysis of wine components, such as metals, specific acids and alcohols, esters, amino acids, and other components (if necessary), we recommend that the winemaker use the services of a wine-specializing analytical laboratory. The analysis covered in this chapter is the basic analysis needed during wine production, from pre-harvest up to bottling. Each analysis contains some background relevant to that analysis and straight procedure on how to run it.

1. Sugar Measurement

The Brix measurement of grapes during the ripening period and at harvest is most important to evaluate the maturity state. This subject has been widely discussed in chapter A. The Brix is a chemical term representing the total soluble solids in a liquid (expressed in grams of solid per 100 grams of solution). In must, these solids contain sugar (90% - 95% of the total solids), organic acids, minerals, proteins, phenols, and many more. There are two methods for Brix measurement: hydrometry and refractometry. Hydrometry is based on the linear dependence of the density of the liquid on its solid's soluble content. The hydrometers are calibrated either directly in Brix units or in density units. If one is using a hydrometer which is density calibrated, figure A.2 (chapter A) can be used to determine the corresponding Brix.

The refractometer measurement is based on the refractive index of the main component in the must - the sugar - and is calibrated to show directly the Brix reading. The use of the refractometer is limited only to must before fermentation, because when alcohol is present, its refraction index interferes with the sugar index.

Temperature corrections have to be made in both methods because the calibration reading is usually set at 20°C. The hydrometry temperature corrections are given in figure A.2, and the refractometry corrections are given in figure R.1.

The precision of both methods is ± (0.1 - 0.2) Brix units.

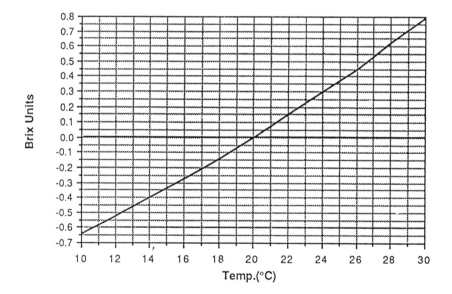

Fig. R.1 - Temperature correction for refractometer measurement of the Brix.

The X axis is the actual temperature of measurement, and the Y axis is the correction to be made to the refractometer reading.

At temperature range between 10-20°C, subtract the Brix correction value, and between 20-30°C, add it.

All corrections are within ± (0.05) Brix accuracy, in the 15-25 Brix range.

Procedure:

Crush the grape sample with a hand-screw crusher and filter the must through double cheesecloth layers, to separate the juice from the heavy lees particles. Avoid hand crushing (which is too soft), or blender crushing (which is too heavy with lees and extracted materials from the skins and seeds). At harvest, if the sample is taken from the stemmer/crusher machine or from the fermentation tank, the must should be filtered only through the double cheesecloth.

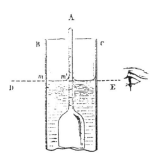

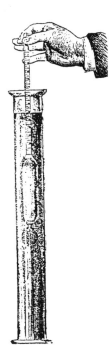

Hydrometer - pour the juice into the cylinder up to the top. Place the hydrometer in the cylinder, while the juice overflows, and rotate it back and forth. This rotation is most important during fermentation, because the CO_2 bubbles stick to the hydrometer surface and change the correct reading. Let it settle and read the liquid level at the lower point of the meniscus. Measure the must temperature and make the necessary correction.

Refractometer - the surface of the refractometer prism should be clean and dry. Place a drop of juice on the surface with a wooden or plastic rod, and close the cover. Look into the eyepiece against direct light and read the Brix (where the sharp boundary between the bright area and dark area meet). Measure the room temperature and make the necessary correction.

Calibration of the refractometer is done from time to time by adjusting its reading at two extreme points - 0°B and 20°B. The 0°B is read by placing a drop of water on the prism. The 20°B is prepared by dissolving 20 grams of dry sucrose in 80 grams of water. The adjustment of the refractometer is usally done with a tiny screw mounted on the refractometer. In general, the refractometer measurements are slighly lower than the hydrometer measurements.

2. Acidity

Total Acidity (TA)

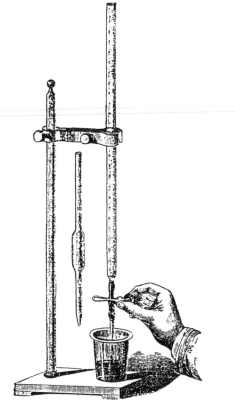

The total titratable acidity includes all organic acids found in the wine, whose origin are either from the grapes (mostly tartaric, malic, citric) or as a result of the fermentation (mainly succinic, acetic, butyric, lactic). The usual range of total acidity in must and wine is in the range of 0.4 to 1.0% expressed as tartaric acid (4 gram - 10 gram/L).

The major changes of acidity during the wine processing occur after acid corrections, after ML-fermentation and after cold stabilization. The acidity is usually expressed as tartaric. It is sometimes expressed as sulfuric acid (in France) or, as acetic acid (when dealing with volatile acidity). The transformation between these acid expressions is based on their equivalent weight ratio:

Acid	Molecular Weight	Equivalent Weight	Factor
Tartaric	150	75	1.00
Sulfuric	98	49	0.65
Acetic	60	60	0.80

Note that the transformation factor for tartaric acid is 1.00. Thus, to convert, for example, the acidity expression from tartaric acid to acetic, multiply the concentration by the transformation factor 0.80. if the reverse is needed, (to calculate e.g. the fixed acidity, see later) multiply the acetic acid concentration by 1.25 (1.00/0.80).

The titration is done with 0.1N NaOH (Sodium Hydroxide). As a result of the titration of weak organic acids with strong base, the neutral salts are basic. The titration curve, followed by pH measurement, will generally appear as a wide S-shape (figure R.2), where the end point is between pH = 8.0-8.5. For wine analysis, the convential end point is at pH = 8.2-8.3, which is the pKa of phenolphthalein indicator. There are, therefore, two methods to determine the titration end point. Either by phenolphthalein indicator (which turns from colorless into pink-red), or by a pH-meter at pH = 8.2. The second method has the advantage in red wines, where it is quite difficult to decide where the end point is, because of the anthocyanin's interference.

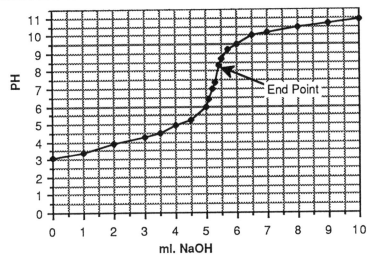

Fig. R.2 - Titration curve of weak organic acid with strong base.

The accuracy of the TA measurement is about 0.1 gram/L.

A major cause of mistaken determination of the TA is the presence of carbon-dioxide in the wine during, and also for a very long time after fermentation (especially if the wine is kept at cool temperatures). Before the TA measurement is done, the CO_2 should be removed. (When unfermented must is analyzed, there is no need for CO_2 removal).

Procedure

Measure exactly 100 ml of wine with a volumetric cylinder and transfer it to a 250 ml Erlenmeyer flask. Bring it to boiling point for a few seconds, and let it cool to room temperature. Pour the wine back into the volumetric cylinder and bring the volume back to 100 ml with distilled water (usually it will be about 90 - 95 ml after the CO_2 removal). Pipette 5 ml of this de-gassed wine into a 50 ml Erlenmeyer, add 20 ml of distilled water and a few drops of 1% phenolphthalein solution. Put a magnetic rod in the Erlenmeyer and place it on magnetic stirrer at a moderate spinning rate. Titrate from a burette with 0.1N NaOH up to the pink color end point. In using a pH-meter instead of indicator (in red wine), titrate until the pH reads 8.2.

The general formula for calculating the percentage of TA as tartaric acid is:

$$\% \ (W/V) = 7.5 \ \frac{Vb \times N}{Vw \times F} \qquad (16)$$

where: Vb = ml of NaOH used in the titration.

Vw = ml of wine sample being titrated

(5 ml in the above procedure).

N = Normality of the NaOH (0.1N).

F = Normality factor in case the NaOH concentration is different from 0.1N.

The NaOH 0.1N solution is prepared by weighing 4 grams of NaOH in a 100 ml beaker. Then add about 50 ml of distilled water to the beaker to dissolve the NaOH. When it has dissolved, transfer the content to 1000 ml volumetric flask. Rinse the beaker 2 to 3 times with water to the volumetric flask and bring to volume (1L). Calibrate the NaOH solution (to determine F) with standard commercial 0.1N HCl (sold in ampules which are diluted to 1L in a volumetric flask). For titration, use any indicator.

The factor value F is given by:

$$F = \frac{N_2 V_2}{N_1 V_1} \qquad (17)$$

where: N_2 = Normality of HCl (0.1N).
V_2 = ml of HCl used in titration.
N_1 = Normality (prepared) of NaOH (0.1N).
V_1 = ml of NaOH taken for titration (generally 10 ml).
(e.g., if 10 ml of NaOH where titrated by 9.8 ml of HCl, then F = 0.98. The exact NaOH concentration is 0.098 N).

Instead of preparing the NaOH solution it is also possible to buy NaOH in ampules which have to be diluted to 1L volume (final concentration of 0.1N) or even to buy ready to use 0.1N NaOH solutions.

Due to the fact that NaOH interacts with the CO_2 in air, the concentration is not stable and the factor F has to be determined from time to time.

If one uses 5 ml of wine for the titration, then:

$$\boxed{\text{Total Acidity (gm/L)} = 1.5 \text{ x Vb x F}} \qquad (18)$$

Volatile Acidity (VA)

The volatile acidity measurement which mainly contains acetic acid serves as an indicator for microbiological spoilage of the wine. Its legal limits in the United States are 1.2 gm/L for white wine and 1.4 gm/L for red. The volatile acidity determination is based on steam distilling of the volatile acids from the wine, and titrating the distillate. A source of mistakes may be the CO_2 and SO_2 presence in the wine, which may be distilled with the other volatile acids. The CO_2 has to be removed from the wine in the same way as for TA measurement. The SO_2 can be determined and its acidic equivalent subtracted from the VA. If sorbic acid is added to the wine, it can also be distilled with the steam. In this case also, its concentration in the distillate has to be determined separately and subtracted. The accuracy of the measurement is 20 mg/L. The steam distillation is done by a simple assembly of equipment as shown in figure R.3 or by commercial cash distillation apparatus.

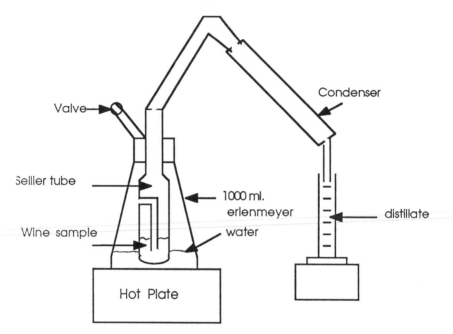

Fig. R.3 - Steam Distillation Apparatus

Procedure:

Pipette 10 ml of de-gassed wine (see TA procedure) into the Sellier tube, connect the tube to the erlenmeyer, filled with 300 to 400 ml distilled water and some boiling stones. Place the erlenmeyer on a hot plate and assemble the condenser and a receiver (100 ml grading tube). Flow cooling water through the condenser and warm the plate at a high level. When the water begins to boil, close the valve on the erlenmeyer, which then will force the steam to flow through the wine inside the Sellier tube into the condenser. Collect the distillate up to 90ml (this value is not exactly essential, but should be repeatable in the series of measurements in order to have a reproducible procedure for the analysis). Before turning off the heater, open the erlenmeyer valve to equalized pressure in the Sellier tube with the atmospheric pressure. Transfer the distillate to a 200 ml erlenmeyer, add a few drops of 1% phenolphthalein, and titrate (as in TA procedure) with NaOH 0.1N. The concentration of the volatile acid expressed as acetic acid is:

$$VA\ (gm/L) = 60\ \frac{Vb \times N}{Vw \times F} \qquad (19)$$

where: Vb = ml of NaOH used in the titration.
$\quad\quad Vw$ = ml of wine sample.
$\quad\quad N$ = Normality of the NaOH (0.1N).
$\quad\quad F$ = Normality factor in case the NaOH concentration is different from 0.1N
In cases where $N = 0.1N$ and the wine sample is 10 ml:

$$VA\ (gm/L) = 0.6 \times Vb \qquad (20)$$

Concerning SO_2 usually the correction is not needed, unless it is at a high concentration and the VA approaches the legal limit. In this case, the SO_2 concentration of the distillate (after finishing the VA titration) can be done by the free SO_2 procedure (see later in this chapter).
The SO_2 acidity equivalent expressed as acetic acid is:

$$SO_2\ (gm/L) = 60\ \frac{Vi \times Ni}{Vw \times F} \qquad (21)$$

where: Vi = ml of iodine solution used for titration.
$\quad\quad Ni$ = Normality of the iodine solution.
$\quad\quad Vw$ = ml of wine sample.
$\quad\quad F$ = Normality factor of iodine solution.
The SO_2 acidity equivalent then, is subtracted from the VA value. For more details see section R.4 on SO_2 analysis.

Fixed Acidity (FA)

The fixed acidity is the total titratable acidity in the wine, minus the volatile ones. Before subtraction, the VA expressed as acetic acid has to be converted to tartaric acid expression by mulitplying by 1.25:

$$\boxed{\text{Fixed acidity} = \text{Total Acidity} - 1.25 \text{ x (Volatile Acidity)}}$$

pH

The pH is one of the most important factors in winemaking, and its measuring starts during the ripening period of the grapes and is followed all the way through the wine processing. Its value is not simply correlated with the acid's concentration in the must or wine, because these acids are weak acids with low pKa's (Table A.1), and also because of the potassium ion present in the must. The usual pH range in must and table wine is between pH = 3.0 to pH = 3.8.

The pH measurement is made by pH-meter, which measures the potential difference due to proton concentration differences between a glass electrode and a calomel reference electrode. Sometimes these two electrodes are combined in one electrode. The pH-meter range is from 0.0 to 14.0 and generally with 0.01 units of accuracy. Before measuring, the pH has to be calibrated with a buffer solution near the measuring value. In measuring wine pH, one can use either a commercial standard buffer solution of pH = 4.00 and pH = 7.00, or a saturated solution of potassium-hydrogen-tartarate (KHT), which has a pH = 3.55 ± 0.01 at 20 to 35°C range. Preparing this solution is simple; add about 1 gram of KHT to 100 ml of distilled water, stir with a magnetic stirrer for a few minutes (not all the KHT will be dissolved) and the saturated solution is ready. The solution is very stable and good for use as long as mold does not grow in it. A fresh solution is recommended from time to time.

One cause for mistakes in the pH measurement can be the existence of CO_2 in the wine, which should be removed before the pH measurement is done.

Procedure:

Warm up the pH-meter

Calibrate the pH-meter with the buffer solutions. Put about 40-50 ml of must or a de-gassed wine sample (see TA measurement) in a 100 ml beaker or erlenmeyer containing a small magnetic rod and place the beaker on a magnetic stirrer. Turn the stirrer on at a low spinning rate, and place the electrode (after rinsing with water) in the beaker, below the liquid surface level (be careful not to break the electrode with the spinning magnetic rod). After the pH reading is stabilized, read the pH.

While not in operation, the electrode should be immersed in distilled water or in a buffer solution.

When the pH reading is unstable and irreproducible, the reason in most cases relates to the electrode surface. In such a case it is possible to clean its surface by placing it in 0.1N HCl for a day, then in 0.1N NaOH for another day and back to HCl for a couple of hours, and lastly in buffer solution at about pH = 7 for an hour.

3. Alcohol

Alcohol is the major product of alcoholic fermentation. Its concentration range is about 7-13.5% in white table wines, and 10-14% in red wines. In fortified and dessert wines it may range up to 24%. The formal expression of alcohol concentration in alcoholic beverages is given by volume of alcohol per volume of liquid (v/v).

The common methods used for determination of alcohol concentration include:

a. by measuring the boiling point of the sample.
b. distillation and measuring the alcohol content by specific density or by refractive index.
c. by chemical methods (oxidation).
d. by gas-chromatography.
e. by enzymatic method.

The most suitable and inexpensive methods for a small winery's laboratory are the boiling point method (a) and the distillation method followed by specific gravity measurement (b).

Boiling Point Method

This method is based on the dependence of the boiling point of a water-alcohol mixture on it's alcohol concentration. The higher the alcohol content, the lower the boiling point. The instrument designed for this measurement is called an ebulliscope or ebulliometer (see figure R.4). Its thermometer scale is calibrated directly with the alcohol concentration, or sometimes the calibration ruler is separate from the mercury scale.

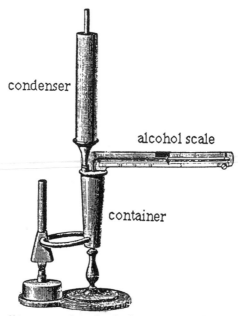

Fig. R.4 - Ebulliometer for alcohol concentration measurement.

The solids extract in the table wine, in principle, can raise the boiling point, although in practice they have virtually no effect on the reading because of their very low molar concentration. Dry extract is in the range of 20-40 gram/L which on molar basis has negligible influence on the boiling point (raising it by 0.52°C/mole solid per liter). In dessert wine with high alcohol and high sugar concentrations, the wine has to be diluted with water to meet the ebulliometer alcohol range (generally 0 to 15%), so the sugar effect is reduced by this dilution. At 10% sugar in the wine, the ebulliometer will show about 0.5% lower alcohol reading than the true one.

The precision range of the boiling point method is about 0.15% of the alcohol reading.

Procedure:

Pour distilled water into the ebulliometer container, up to the lower mark line (below the thermometer's mercury ball). Light the lamp under the container and let the water boil. The steam coming up will warm the thermometer mercury ball up to the water's boiling temperature. When the mercury at the capillary tube comes to rest, adjust the calibration scale above the mercury capillary tube to zero, and tighten it with the screw. This is the atmospheric pressure adjustment of the boiling point for 0% alcohol. Pour off the water, wash the container with the wine under study, and pour in the wine sample, up to the second marked line (above the mercury ball). Fill the condenser with water and place it above the container. Light the lamp, and when the mercury come to rest, record the alcohol reading on the scale.

Distillation Method

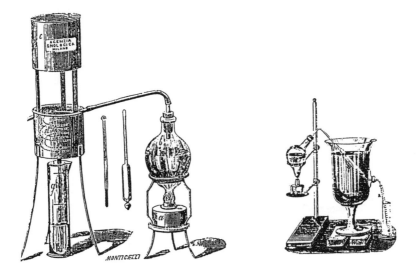

Fig. R.5 - Distillation apparatus.

This method is based on distilling out the alcohol from the wine sample (figure R.5), then bringing the distillate back to the original volume of the sample, and measuring the alcohol concentration by the specific density of the water-alcohol mixture. This measurement can be done either by hydrometer or by picnometer. The alcohol concentration is then found by standard tables or graphs presenting the dependence of the density of the water/alcohol mixture on the alcohol concentration (figure R.6).

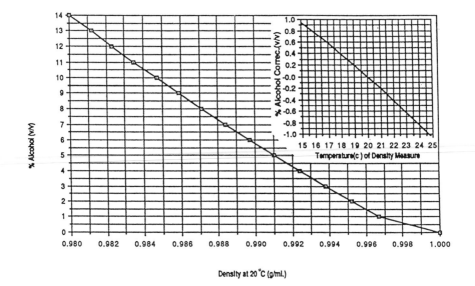

Fig. R.6. -The density of water-alcohol mixture at 20°C.

The temperature correction is shown in the inner graph for the range of 15-25°C, for wines at an alcohol content range of 10-15%. At the above ranges of temperatures and alcohol content, the corrected deviations are in the range of ±0.1% alcohol.

Procedure:
Put about 300 ml of the wine to be tested in a thermostated bath at 20°C. When equilibrium has been reached, measure exactly 250 ml of wine in a volumetric flask. Transfer the wine into the distillation container and rinse the volumetric flask, with 50 ml of water into the distillation container. Add some boiling stones, and place the volumetric flask under the condenser. Run water in the condenser, and begin to distill the wine at a moderate rate, until about 220 ml of distillate is collected. Place the volumetric flask in the bath at 20°C and when temperature equilibrium is reached, add distilled water to the original volume (250 ml). Save the residue left in the distillation flask for the dry extraction analysis (see later). To measure the density of the water/alcohol mixture, pour the distillate into a cylinder and place in the alcohol hydrometer. Read the density (or the percentage of alcohol on the hydrometer scale, and check the temperature of the liquid. If needed, make a temperature correction. The precision of this method is 0.1% alcohol (at 20°C).

4. Sulfur Dioxide

a. White Wine
The determination of SO_2 is based on its reduction potential, and is done by iodine as an oxidation agent (Ripper method):

$$SO_2 + I_2 + 2\ H_2O \text{ ------> } H_2SO_4 + 2HI \qquad (22)$$

The free SO_2 in wine is titrated with iodine in acidified media in order to minimize the reduction potential of other components. The reaction takes place only with the free SO_2 present in the wine, but not with the bound SO_2 which does not react directly with iodine.

The total SO_2 (bound and free) is measured by first hydrolysing all of the bound SO_2 by a strong base media. and then titrating the sulfure dioxide as for free SO_2.

The titration is done with iodine solution 0.025N, and starch solution as an end point indicator.

To prepare the iodine solution; weigh 3.25 gram of iodine and 6.25 gram of KI in a beaker. Add 100 ml of water and mix well. Transfer to a 1,000 ml volumetric flask and bring to volume (1L).

Keep in a brown colored bottle and in darkness when not in use. To determine the normality factor F of the iodine solution, calibrate it with a 0.02N thiosulfate solution. The calibration is done by titrating the assumed 0.025N iodine solution with 0.02N thiosulfate, using starch as an indicator.

The reaction is:

$$2Na_2S_2O_3 + I_2 ------> 2NaI + Na_2S_4O_6 \qquad (23)$$

To calibrate: Pipette 10 ml of the iodine solution into a 50 ml erlenmeyer, and titrate with thiosulfate until the iodine red color almost disappears. Add 5 ml of starch to the iodine solution (it will become blue), and continue to titrate until the blue color vanishes. The normality factor F is then given by:

$$F = \frac{Ns \times Vs}{N_I \times V_I} \qquad (24)$$

where: Ns = Normality of the thiosulfate (0.02N)

Vs = ml of thiosulfate used in titration.

N_I = Normality (prepared) of iodine.

V_I = ml of iodine sample (10 ml).

(e.g. if in the above case, 9.65 ml of thiosulfate have been used, then F = 0.772).

To prepare a 0.02N thiosulfate solution, weigh 4.962 grams of sodium-thiosulfate pentahydrate ($Na_2S_2O_3 .5H_2O$) in a beaker, dissolve with some water and transfer to a 1,000 ml volumetric flask. Rinse the beaker twice with water into the flask and bring to volume (1L).

The starch solution is made by mixing about 20 grams of starch in 500 ml of water and boiling the mixture for a few minutes. The solution may become moldy after some time, so a fresh solution has to be made occasionally. The accuracy of the method is ±5% of the measured value.

Procedure:

Free SO$_2$

Pipette 50 ml of the wine into a 150 ml erlenmeyer. Add approximately 10 ml of diluted H_2SO_4 (25%), and 10 ml of starch solution. Put in a magnetic rod and place on the magnetic stirrer at a low spinning rate. Titrate with the iodine solution, up to the end point of the starch (blue), which does not vanish for half a minute.

From the relation of Eq.22:

$$SO_2 \text{ (mg/L)} = \frac{M \times N_I \times V_I \times F \times 1000}{Vw \times 2} = \frac{64 \times N_I \times V_I \times F \times 1000}{Vw \times 2} \quad (25)$$

where: M = molecular weight of SO_2 (64)
$\quad N_I$ = Normality of the iodine solution
$\quad F$ = the factor of the iodine solution (Eq.24)
$\quad V_I$ = volume of the iodine (ml) used in the titration
$\quad Vw$ = volume of wine in the analysis
for Vw = 50 ml, and N_I = 0.025N:

$$SO_2 \text{ (ppm)} = \frac{64 \times 0.025 \times 1000 \times V_I \times F}{50 \times 2} = 16 \times V_I \times F \quad (26)$$

Total SO$_2$

Pipette 25 ml of wine into a 150 ml erlenmeyer and add 25 ml of 1N NaOH. Put in a magnetic rod, place on a magnetic stirrer, and stir at a slow rate for 15 minutes. Add 15 ml of diluted H_2SO_4 (25%) and 10 ml of starch. Titrate with iodine solution as with free SO_2. The concentration of the total SO_2 is given (for N_I = 0.025N, Vw = 25 ml) by:

$$SO_2 \text{ (ppm)} = 32 \times V_I \times F \quad (27)$$

b. Red Wine

Using the iodine method to determine the SO_2 concentration in red wines causes some difficulties in observing the end point of titration because of the strong red color. It is partially possible to overcome this difficulty by diluting the wine sample with water and to carefully follow the color changes during titration. If one wants to use the iodine method, the procedure is: Pipette 50 ml of wine sample into 500 ml erlenmeyer and add 100 ml of water, 20 ml of diluted H_2SO_4 (25%), and 20 ml of starch. Titrate with the iodine solution and watch the color carefully. The solution color is kind of a mixture of red and blue, which at the end point loses its red color. Some experience is needed to carry on this titration. Side light on the titrated erlenmeyer helps very much.

A no less serious problem with the Ripper method in red wines is that the tannins and color compounds utilize certain quantities of iodine, which make the results seem higher than they really are (as much as 20 ppm higher). This may also happen in wines containing high concentration of aldehydes such as botrytis late harvest wines.

Another method for measuring sulfur-dioxide which by-passes these two difficulties is based on removing the SO_2 from the sample by air stream, and collecting it in an oxydizing solution (H_2O_2) according to the reaction:

$$SO_2 + H_2O_2 \text{------} > 2H^+ + SO_4^= \tag{28}$$

which is then titrated by sodium-hydroxide to determine the concentration of the acid being formed. This method is good as well for white wine, especially if ascorbic acid has been added. In such a case the Ripper method will show a higher SO_2 level than actually exists because of the redox capacity of the free ascorbic acid left in the wine.

This method is called the aeration-oxidation method. The apparatus needed for this measurement is shown in figure R.7.

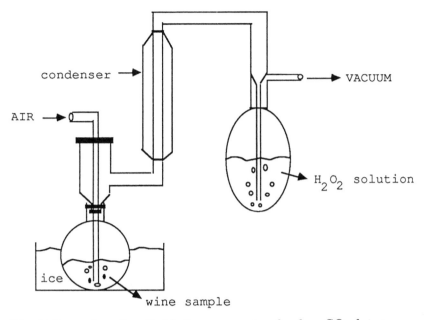

Fig. R.7 - Aeration-Oxidation apparatus for free SO_2 determination in wine.

The wine sample is placed in a glass bulb (100 ml of volume) and the air stream flows into the sample by a vacuum exerted at the other side of the apparatus. Hydrogen-peroxide (0.5%) and indicator is placed in the other bulb where the stream of air carrying the SO_2 passes through it. The condenser function is to limit volatile organic acids to pass through. The wine sample is cooled in an ice bath to minimize dissociation of bound SO_2 at such low temperature. The accuracy of the method for free SO_2 measurement is ±2 ppm.

Procedure:

Free SO_2
Pipette 20 ml of wine into the glass bulb and add 20 ml of phosphoric acid (25%). Place it in an ice bath, connect the other parts of the apparatus, and add to the collecting bulb 20 ml of 0.5% H_2O_2 solution.

The H_2O_2 0.5% solution should be prepared from 30% solution by diluting 1.6 ml to 100 ml with water. The 30% H_2O_2 should be kept in the refrigerator, and the 0.5% should be prepared daily.

Add to the hydrogen-peroxide solution in the collecting bulb a few drops of methyl red indicator, and one drop of phosphoric acid (25%). The solution will turn purple-red. Titrate this blank solution with 0.01N NaOH (few drops) to color change (pale-yellow). Connect the bulb containing the solution to the apparatus, and open the vacuum line (water pump). Start the cold water stream in the condenser and let the air flow through the system for 10 minutes. The H_2O_2 solution will turn red. Remove the bulb with the H_2O_2 solution and titrate it with 0.01N NaOH till the color changes to yellow. The calculation for the above procedure, where Vw = 20 ml and Nb = 0.01N:

$$SO_2 \text{ (ppm)} = 16 \text{ x Vb x F} \qquad (29)$$

where: Vb = the volume (ml) of the base used in the titration.
  F = the NaOH factor (which should be calibrated frequently because of the low concentration of the solution).

Total SO_2

The same procedure as for free SO_2 except for two modifications; the wine sample, instead of being cooled with ice, is heated to boiling point, while the aeration takes place and the air flow is expanded to 30 minutes. The bound SO_2 is dissociated and is removed by the air stream as the free SO_2. The calculation is the same.

The free and the bound SO_2 concentration can be measured in a raw on the same wine sample. Measure first the free SO_2 by cooling the sample in ice bath, and after titrating the H_2O_2 red solution to the yellow end-point, place it back in the system, and replace the ice bath with heat source. The second titration will evaluate the bound SO_2. The sum of the two titrations (free and bound) is the total SO_2 concentration.

5. Residual Sugar

The term residual sugar in table wines refers to a range of 0.2% to about 3% (2-30 gm/L) of sugar. If more sugar is present in the wine, the wine is not considered to be a dry wine, and the determination of its sugar content is different. If the fermentation has been run to "dryness" there is still about 0.2 to 0.3% of unfermentable pentose sugars. Any sugar level above that is either a residual sugar, caused by stopping the fermentation (by itself or on purpose), or has been added to the wine after fermentation.

The analytical principle of sugar determination is based on its reducing potential, where the oxidation agents generally used for the analysis may be copper ion (Cu^{++}), or enzymatic reagents. In this reaction the reducing sugars are oxidized to acids. The reaction is not specific for sugar only and any other reducing substance present in the wine may react at the reduction-oxidation media of the test. Still, the reliability of the test is based on the fact that the major reducing components in wine, at the test conditions, are the reducing sugars. From the many methods for residual sugar determination, we shall contend here with two of the easiest and fastest methods which have good accuracy in respect to practical wine analysis.

a. Quick Tablets Reagent Method

This method is generally used to test the reducing sugar con-centration in the urine of diabetics. It is very simple, very quick and sensitive to 0.1% sugar in the range of 0 - 1% sugar concentration. The principle of the test is a reaction of the sugar with copper ion to produce different colored solutions from blue to brown, depending on the sugar concentration. The determination of the sugar per-centage is made by comparing the test solution with a set of given colors. The test kit is commercial, and is available at any pharmacy. The exact directions for use are given in the kit.

b. H. Rebelein Method

This is a fast method compared to other traditional methods, but much slower than the tablet method. Its principle is based on reducing a quantitative amount of copper ion by the sugar, where the copper ion is in excess. Then, adding an iodine anion in excess, reducing the extra copper ion left over, and then finally determining the free iodine by thiosulfate solution. The reactions are:

$$3I^- + 2Cu^{++} ------> I_3^- + 2Cu^+$$

(30)

$$I_3^- + 2S_2O_3^= ------> 3I^- + S_4O_6^=$$

The copper ion in the reagent is complexed by tartarate anion in the basic media of the reaction. By using given volumes of calibrated standard solutions of $CuSO_4$, KI, and $Na_2S_2O_3$ the measured volume of thiosulfate which is used to react with the excess free iodine will determine the sugar concentration.

Preparing the standard solutions:

a. Copper Sulfate - Weigh 41.92 grams of $CuSO_4.5H_2O$, add 10 ml of H_2SO_4 1N, dissolve in a beaker, transfer to a 1000 ml volumtric flask, and bring to volume (1L) with distilled water.

b. Tartarate solution (basic) - Weigh 230 grams of sodium potassium tartarate + 80 gram NaOH, dissolve and bring to volume with distilled water in a 1L volumetric flask.

c. KI solution - Weigh 300 gram KI, add 100 ml NaOH 1N, and bring to volume with distilled water in a 1L volumetric flask.

d. Thiosulfate - Weigh 13.77 gram $Na_2S_2O_3.5H_2O$ and add 50 ml NaOH 1N. Dissolve in distilled water and bring to volume in a 1L volumetric flask.

Procedure:

In a 100 to 150 ml erlenmeyer, put 10 ml of copper sulfate solution + 5 ml of tartarate solution + some boiling stones. Pipette 2 ml of the tested wine, bring quickly to boiling point and keep there for 90 seconds. Cool the solution quickly and add 10 ml of the KI solution + 10 ml of starch + 10 ml of H_2SO_4 (15%) and a magnetic rod. The solution will have a blue color. Place on a magnetic stirrer and titrate with the thiosulfate solution until the blue color disappears. Record the ml used in the titration. Repeat exactly the above procedure with 2 ml of water (as blank), instead of the wine sample, and record the thiosulfate ml used in the blank titration. In this procedure, each ml of the difference between the wine sample and blank is equivalent to 1 gram/L of reducing sugar. The method is good for samples within the range of 0 to 25 gram/L of sugar.

If the sugar concentration in the wine is higher, dilute the wine with water and multiply the result by the dilution factor. The sign that the sugar concentration is too high for the test is that after adding the acidic KI solution and the starch, the solution does not turn blue.

For red wines the end point is difficult to determine, so it is recommended to decolor the wine sample with activated carbon before the test. The decoloration should be done with a minimum amount of carbon.

For dessert wines with high sugar levels, the determination can be done either by dilution of the sample to the residual sugar range that can be tested by the above methods, or the sugar content can be determined by the dry extract method (see later). The no-sugar dry extract in wine ranges between 10-30 gram/L with an average of 20 gram/L. For high sugar wines, the sugar level can be measured quite satisfactorily by determining the total dry extract and subtracting 20 gram/L of the non-sugar extract from the total one. The difference represents the sugar content in the wine.

6. Dry Extract

The dry extract in wine contains sugar and all non-volatile compounds such as acids, tannins, esters, minerals and others. The concentration of the non-sugar extract in table wines is between 10-30 gram/L. In late harvest wines it is at the 30 gram/L and above. Addition of water and sugar, for example, to the must can be detected by the non-sugar dry extract concentration. In California, the minimum level of non-sugar dry extract is 17 gram/L for white wines, and 18 gram/L for red. Too low of a dry extract figure in wine indicates a possibility of water dilution.

There are several methods for determining the total dry extract. Some of them are based on drying the wine and weighing the extract. The most accurate method is to dealcoholize the wine by evaporation, then to bring it back to the original volume with water, and measuring the density of the solution by hydrometer. The concentration of the dissolved dry extract is found from density/concentration tables or graphs (figure R.8).

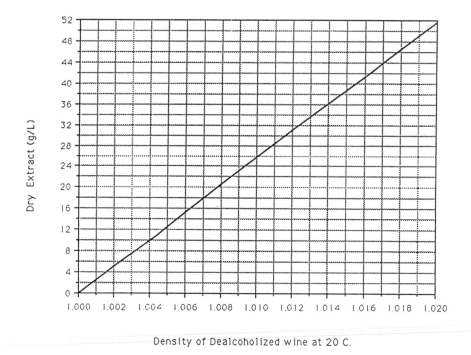

Density of Dealcoholized wine at 20 C.

Fig. R.8. -The density of dealcoholized wine at 20°C for dry extract determination.
(assuming the density of water at 20°C is 1,000).

Procedure:

The residue left over from the alcohol analysis is suitable for the dry extract measurement (section R.3, distillation method procedure). Transfer the residue solution into a 250 ml volumetric flask, mix well and rinse the residue twice with water into the flask. Add water below the 250 ml mark and place in a 20°C bath. When temperature equilibrium has been reached, bring to volume.

If there is no residue from alcohol analysis, measure 250 ml of wine in a 250 ml volumetric flask placed in a 20°C bath, transfer to a 500 ml beaker and rinse twice with 25 ml of water. Put boiling stones in the beaker and evaporate until about 30-40 ml of liquid is left. Then proceed by bringing back to volume, as was described above.

Pour the dealcoholized wine into a cylinder and insert hydrometer which is in the 1.000 to 1.020 density range. Use figure R.8 to determine the dry extract concentration.

7. Tannins and Pigments

The measurement of tannins in wine gives the winemaker a tool to compare the phenolic components in different wines (or must), or to follow the changes in the wine after certain treatments, such as skin contact, fining, stabilizing and aging.

The range of the total phenolic compounds (tannins and pigments) is 100 to 1000 mg/L in white wines, and 300 to 3000 mg/L in red wines. Average typical values are close to 300 mg/L in white, and 1500 mg/L in red. The method presented here for determining the total phenolic content is based on the oxidation of the total oxidizable components in dealcoholized wine, and then, after removing the tannin and color pigments from the wine (by activated carbon), a second oxidation takes place. The difference between the two measurements is related to the tannins and pigments in the wine. The oxidation agent is a solution of 0.1N potassium-permanganate ($KMnO_4$). Its volume needed for the oxidation is determined by titration, using indigo-carmine as an indicator.

To prepare the permanganante 0.1N solution: Weigh 3.2 grams of $KMnO_4$ and dissolve in a beaker with about 200 ml of distilled water. Transfer to a 1000 ml volumetric flask, rinse the beaker into the flask, and bring it to volume (1L). Leave for a couple of hours and then filter the liquid into a brown bottle.

Calibrate the exact normality with sodium-oxalate according to the reaction:

$$2KMnO_4 + 5Na_2(COO)_2 + 8H_2SO_4 \text{------}> 2MnSO_4 + 5Na_2SO_4 + K_2SO_4 + 8H_2O + 10CO_2$$

$$(31)$$

The equivalent weight of the oxalate in this reaction is its molecular weight/2 and that of the permanganate is its molecular weight/5.

Weigh 200 mg of sodium-oxalate in a 500 ml beaker, add 200 ml H_2SO_4 1N and a magnetic rod. Place on a magnetic stirrer plate, and heat the solution to about 80°C and stir. When hot, titrate this solution with the permanganate solution to a pink permanganate color, which stays for at least half a minute.

The exact normality of the permanganate is:

$$\text{permanganate (N)} = \frac{1000 \text{ x G}}{67 \text{ x V}} = 14.92 \text{ G/V} \qquad (32)$$

where: G = weight of sodium-oxalate (grams).

V = ml of titrated permanganate.

For the above quantities of $KMnO_4$ and $Na_2(COO)_2$, about 30 ml of permanganate should be needed for the titration.

To prepare the indigo-carmine indicator solution, dissolve 3 grams of indigo-carmine in 500 ml of water and add 50 ml of concentrated H_2SO_4. After the indigo is dissolved, filter and store.

Procedure:

Pipette 10 ml of dealcoholized wine (after it has been brought back to the original volume) into 1L beaker. For details on how to dealcoholize the wine, see the procedure of the dry extract measurement in section R.6. Add to the beaker 500 ml of water, 10 ml of indigo-carmine indicator, and a magnetic rod. Place on a magnetic stirrer at a high spinning rate. Titrate with the permanganate solution to yellow end-point. Record the ml titrate (a). To about 30 ml of the same dealcoholized wine, add 1-2 grams activated carbon, mix well, and filter. Add another portion of carbon to the filtrate and filter again. If the wine is still colored, add one more time. If after the filtration, the wine is black from carbon particles, filter again. Pipette 10 ml of the filtered detanninized wine into a 1L beaker and follow the above procedure. Record the ml titrated (b). The difference (a-b) is related to the total tannin and pigments in the wine. By an average experimental acceptance, the concentration of the phenolic components is:

$$\text{Total Phenolic (mg/L)} = 4.16 \text{ x } 10^4 \frac{\text{Vp x N}}{\text{Vw}} \qquad (33)$$

where: Vp = (a-b) ml of permanganate titrated.

Vw = ml of wine sample.

N = Normality of the permanganate.

For the above volume of wine sample (10 ml):

$$\boxed{\text{Total Phenolic (mg/L)} = 4160 \times N \times (a\text{-}b)} \qquad (34)$$

The repeatability of the method is ±5%.

8. Ammonia

The ammonia determination in the must prior to fermentation is important in order to know if there might be a deficiency in the nitrogen nutrient for the yeast growth. The general concentration range is between 50 - 200 mg/L of NH_3. Content of 150 - 200 mg/L will assure the winemaker that there will be no nitrogen deficiency problem during fermentation. The difference between 200 mg/L and the actual ammonia content in the must can be compensated by addition of di-ammonium-phosphate $(NH_4)_2 HPO_4$ into the must at the beginning of fermentation.

The determination is based on the evaporation of the ammonia gas from an alkaline solution of its salts. When must is analyzed, the sugar in it may release under strong alkaline media some volatile acids, which will partially neutralize the ammonia and consequently, show less ammonia than acutally exists in the must. Therefore, the analysis is done with weak alkaline media by using magnesium oxide. The evaporated ammonia gas is collected into a very weak boric acid in which then, the free ammonia is titrated by hydrochloric acid, using methyl-red as indicator.

The methyl-red solution is made by dissolving 0.25 grams of methyl-red in 100 ml of a 60% alcohol-water mixture.

Procedure:

Pipette 200 ml of must (double cheesecloth filtered), into the boiling erlenmeyer (see figure R.9.), and add freshly prepared solution of 5 grams of magnesium-oxide (MgO) in 50 ml of water.

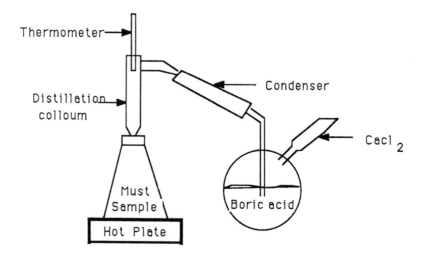

Fig. R.9 - Ammonia Determination Apparatus

Add boiling stones and place on a hot plate. Put 250 ml of 4% solution of boric acid in the receiving container, and assemble the distillation apparatus. Connect the calcium-chloride tube to the boric acid container, and start to distill and collect about 200 ml of distillate. Transfer the ammonia-boric acid distillate solution into a beaker, add a few drops of methyl-red indicator, and titrate with 0.1N HCl to a red end point.

The concentration of ammonia (in mg/L) is:

$$\text{ammonia (mg/L)} = 17000 \ \frac{Na \times Va}{Vw}$$

where: Na = Normality of the HCl.
    Va = ml of titrated HCl.
    Vw = ml of the must sample.
For the above procedure (200 ml of must and 0.1N HCl):

$$\text{Ammonia (mg/L)} = 8.5 \times Va \tag{35}$$

9. ML Determination by Paper Chromatography

This test is important during the ML-fermentation, to enable the winemaker to follow the fermentation rate and to know when it is finished.

The chromatographic test is very simple and it gives qualitative information with regard to the ML-fermentation status.

To prepare the elute solution:

100 ml of n-butanol.

100 ml of distilled water.

15 ml of 1% solution of bromocresol-green indicator (in water).

Put the mixture of the above reagents in separator funnel, shake well and let the two phases separate. Discard the water phase (lower) and leave for a day in the funnel. The next day discard again the water phase and transfer the organic phase to a one gallon jar with wide opening. The height of the solution in the jar should be about 1-2 cm.

The solution should be placed back into the separator funnel after two-three days of use, to separate the excess water. The solution is good for about one week.

Prepare also reference solutions of malic and lactic acids (0.3%) by dissolving 300 mg of each in 100 ml of water.

Procedure:

1. Take chromatograph paper (Whatman #1) and mark a line with a lead pencil about 3 cm from the bottom of one side. On the line mark dots from left to right at about one inch intervals.

Write, in pencil, M for malic under the left dot, L for lactic under the second dot, and continue with the other dots to mark the sample numbers.

2. By using capillary tubes which had been placed for a second in the tested samples, touch the dots on the paper with the proper samples and references. The touching should be very gentle and for a very short time so as not to spread the wet spot on the paper.

Leave to dry for 15 minutes and repeat two-three more times.

3. Fold the paper without overlapping the dots and hold it as a cylinder by paper clips. Place the paper cylinder in the jar and cover

the top. The liquid will start to travel upward on the paper. Wait until the front line of the climbing solvent is about 4 cm from the top (5 - 8 hours).

4. Remove the paper from the jar, open it and hang it to dry. Do not touch the paper except at the edges. Color will develop in about 2 - 3 hours.

5. The background of the paper will be blue, where the white-yellow spots along the paper indicate the organic acids.

The travelling distance of the spots from the base line are in the following order: tartaric acid < citric acid < malic acid < succinic acid.

The absence of the malic acid spot on the chromatograph indicates that the sample has fully undergone ML-fermentation.

10. Microscopic Test

The microscope is widely used in the winery laboratory as a tool to watch and count small particles in the winemaking process, such as yeast, bacteria, fining particles, lees particles, crystals and any micro-particles which may be in the wine. The microscope is used during any stage of the wine processing, from fermentation through racking, fining, filtering and finally bottling.

The microscope ranges of magnification power should be at three approximate values: x100, x400, x1000. For magnification over x1000, an oil immersion technique has to be used. The size of mature yeast cells ranges between 1 to 2 microns (1 micron = 10^{-3} mm). ML bacteria size is about 1.0 microns. The most useful magnification powers for biological tests in the wine laboratory are x500 and x1000.

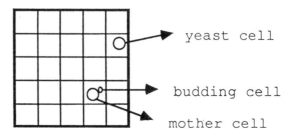

yeast cell

budding cell

mother cell

For non-counting usage, a regular slide glass is used to hold the tested sample. For counting purposes, a special slide with a micro grid on it should be used (figure R.10). This grid is usually 5 x 5 squares, where the square's dimensions are 5×10^{-3} cm, and the distance between the slide surface and the cover glass on the sample is 10^{-2} cm. The view under the microscope contains the yeast cells enclosed in the grid's boundaries.

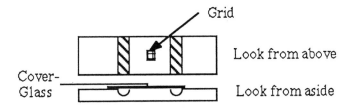

Fig. R.10 - Grid slide for microscopic counting.

One can count the number of the yeast cells in all the sub-squares of the grid, or count some specific representing sub-squares such as the diagonal or any other configuration.

The volume enclosed in each square is, 25×10^{-8} ml, and when counting, for example, yeast numbers, it would be:

$$\text{yeast number/ml} = \frac{N}{n \times V} \qquad (36)$$

where: N = Number of yeast cells counted in n squares.

n = Number of squares in the grid where yeasts are counted, (e.g. those on the diagonal of the 5 x 5 grid).

V = Volume of one square in cm^3.

When the sample contains a very small amount of particles (e.g. during bottling or after filtering), a measured volume of the wine is centrifuged in a small laboratory centrifuge for about 5 minutes. The liquid above the bottom of the centrifuge tubes is decanted off and a small measured volume of distilled water (about 10 times smaller than the original before centrifugation) is added to the residue at the bottom, mixed well and tested under the microscope. The number of particles or cells counted are then multiplied by the concentration factor.

To recognize the dead yeast cells, one can add to the tested solution a drop of methylene-blue a few minutes before testing. The dead cells will be colored while the viable ones will remain un-stained.

EDITOR'S NOTE: *At the time of publication, the use of di-ammonium phosphate $(NH_4)_2 HPO_4$ in wine making is controversial due to the formation of urea, a precurser of ethyl carbamate, a known carcinogen.*

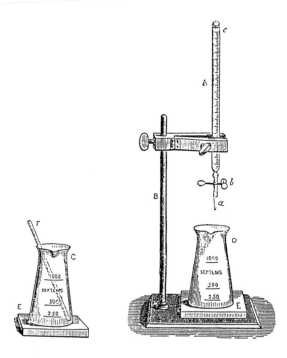

Appendix A - Chemical formulas of compounds related to wine and winemaking

In this appendix one may find the chemical formulas of the major compounds mentioned in the book.

It may be useful for the winemaker to have these formulas at hand in the book, rather than have to refer to other text books.

Along with the formulas are the molecular weight of the compounds and their relevance to winemaking.

The phenolic compounds relating to wine are not included here as they are given in Appendix B.

compound	formula	molecular weight	remarks
acetic acid	CH_3-$COOH$	60.1	By-product of fermentation. Main product of acetobacter spoilage.
acetaldehyde	CH_3-CHO	44	By-product of fermentation.
ascorbic acid	$O=C\diagup\!\!\!\!\overset{O}{}\diagdown\overset{OH}{CH\text{-}CH\text{-}CH_2OH}$ $C===C$ $\overset{}{OH}\quad\overset{}{OH}$	176.1	Vitamin C. Added to wine to inhibit oxidation
citric acid	CH_2-$COOH$ HO-C- - -$COOH$ CH_2-$COOH$	192.1	One of the main acids in wine.
diacetyl	CH_3-CO-CO-CH_3	86.1	By-product of ML fermentation Buttery flavor.
di-ammonium phosphate	$(NH_4)_2\,HPO_4$	132	Ammonium addition to must for yeast growth.
fructose	$CH(OH)$-$CH(OH)$-$CH(OH)$ $CH_2OH \qquad\quad C=O$ CH_2OH	180.1	One of the main sugars in grapes must.

fumaric acid	HOOC-CH=CH-COOH	116.1	Additive to inhibit ML-ferm. Max. limit 3 gm/L
glucose	CH_2OH CH----O $CH(OH)$ $CH(OH)$ CH----CH OH OH	180.1	One of the main sugars in grapes must.
glycerol	$CH_2(OH)$-$CH(OH)$-CH_2OH	92.1	By-product of fermentation
lactic acid	CH_3-$CH(OH)$-COOH	90.1	Main product of ML fermentation.
malic acid	HOOC-$CH(OH)$-CH_2 $COOH$	134.1	One of the main acids in grapes must.
potassium meta-bisulfite	$K_2S_2O_5$	222.3	SO_2 additive in winemaking.
pyruvic acid	CH_3-CO-COOH	88.1	By-product of fermentation.
sorbic acid	CH_3-CH=CH-CH=CH $COOH$	112.1	Additive to wine to inhibit yeast growth. Max. limit 1 gm/L
succinic acid	HOOC-CH_2-CH_2-COOH	118.1	By-product of fermentation.
tartaric acid	$CH(OH)$-COOH $CH(OH)$-COOH	150.1	One of the main acids in grapes must.

Appendix B - Phenolic Compounds (pigments and tannins) in Wine

Phenolic compounds play a very important part in winemaking, as pigments and tannin agents. There are several thousand known to be found in plants, and several hundred found in grapes and wines. In order to have some familiarity in this important group of compounds, the basic structure and chemical formulae will be presented in this appendix.

i. - The monomeric compounds in this category, which are related to wines, can be divided into three major groups, according to their basic structure:

    a. Benzoic acid derivatives

    b. Cinnamic acid derivatives

    c. Flavonoid derivatives.

a. Benzoic acid derivatives $(C_6\text{-}C_1)$

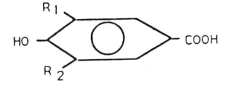

compound	R_1	R_2
p-Hydroxybenzoic acid	H	H
Vanillic acid	H	OCH_3
Gallic acid	OH	OH
Syringic acid	OCH_3	OCH_3

b. Cinnamic acid derivatives (C_6-C_3)

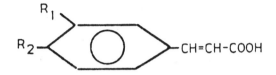

compound	R_1	R_2
Cinnamic acid	H	H
p-Coumaric acid	H	OH
Caffeic acid	OH	OH
Ferulic acid	OCH_3	OH

C. Flavonoid derivatives (C_6-C_3-C_6)

These compounds are based on the flavonoid skeleton which is a three rings molecule, two of them aromatic connected by heterocyclic center ring.

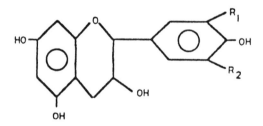

To be familiar with the many flavonoid skeleton compounds, here are the sub-flavonoid groups according to the modifications on the flavan's center ring:

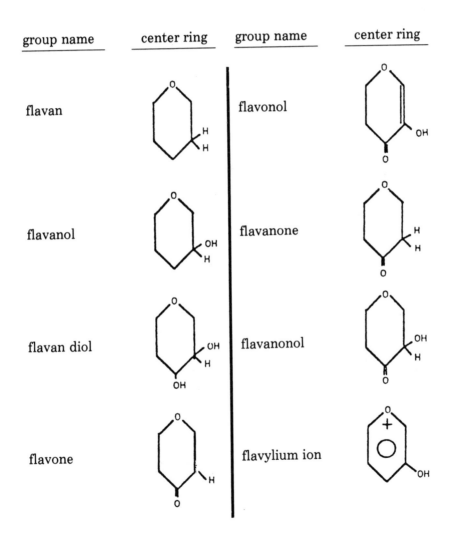

group name	center ring	group name	center ring
flavan		flavonol	
flavanol		flavanone	
flavan diol		flavanonol	
flavone		flavylium ion	

The main groups and related compounds are shown below:

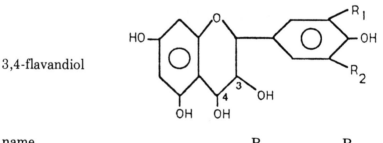

3-flavanol

name	R_1	R_2	
afzelechin	H	H	
catechin	OH	H --->d(+)	
epicatechin	OH	H --->l(-)	
gallocatechin	OH	OH	

The flavanol compounds have bitter taste.

3,4-flavandiol

name	R_1	R_2
leucopelargonodin	H	H
leucocyanidin	OH	H
leucodelphinidin	OH	OH

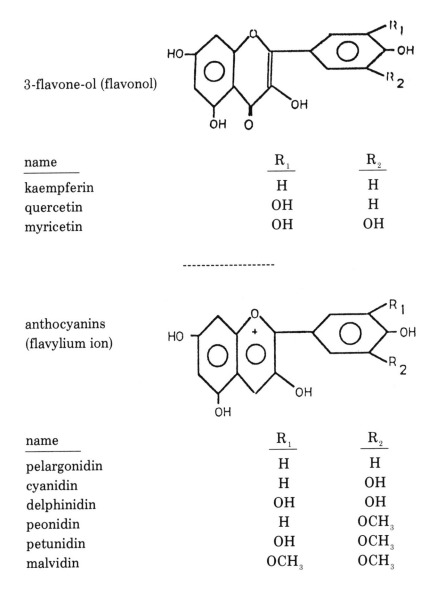

3-flavone-ol (flavonol)

name	R_1	R_2
kaempferin	H	H
quercetin	OH	H
myricetin	OH	OH

----

anthocyanins
(flavylium ion)

name	R_1	R_2
pelargonidin	H	H
cyanidin	H	OH
delphinidin	OH	OH
peonidin	H	OCH_3
petunidin	OH	OCH_3
malvidin	OCH_3	OCH_3

----

ii. - The polymeric compounds (tannins) are structured from the phenolic monomers to form polymers of different length.

a. Condensed tannins

C-C bond or C-O bond of flavonoids (e.g. catechin, epicatechin and leucocyanidin) to form polymers which contain 2-10 monomeric units.

Lignin is a macropolymer mainly of cinnamic acid derivatives monomers.

A typical trimer of catechin is shown here:

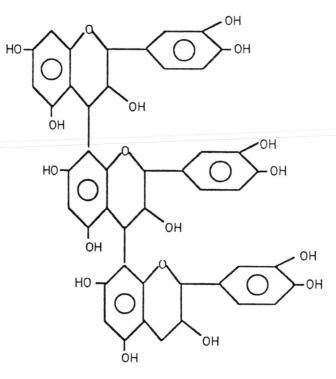

b. Hydrolyzable tannins

Mainly of gallic acid or ellagic acid with sugar molecule.
The major source of these tannins in wine comes from oak extraction.

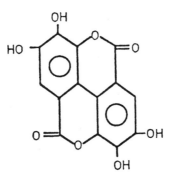

ellagic acid

A typical polymer of gallic acid is shown here:

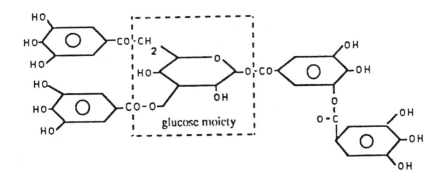

Tannic acid is a polymer of about 8 monomers of gallic acid and glucose. The hydrolyzable tannins can hydrolyze to glucose and gallic or ellagic acids units.

iii. - Combined phenolic compounds contain different combinations of molecules which are present in the wine medium, connected mainly through esteric or etheric bonds.
For example cinnamic acid derivative + tartaric acid

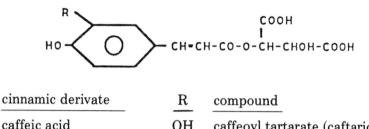

cinnamic derivate	R	compound
caffeic acid	OH	caffeoyl tartarate (caftaric acid)
p-coumaric acid	H	coumaroyl tartarate (coutaric acid)

Another example is flavonoid + glucose + coumaric acid

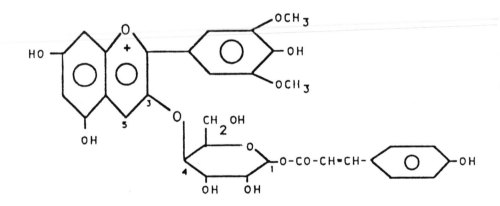

3-malvidin 4-glucose 1-coumarate

One more glucose molecule can be attached at 5 position to become malvidin di-glucoside (found in V.Labrusca species).

When anthocyanins are bound to sugar molecules they are called anthocyanidins.

iv. - **Red grapes color**

The major component in grapes color is the flavylium ion, whose color and intensity are pH dependent.

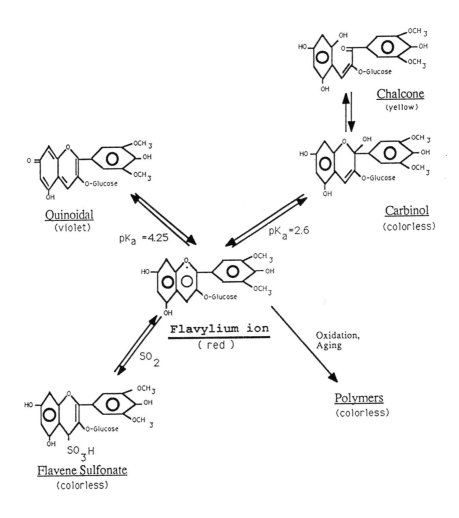

Figure AppB.1 - Falvylium ion equilibrium in red wine.

The concentration of the red flavylium ion (F^+) is pH dependent between all equilibrium species in solution.

According to the basic principles of chemical equilibrium:

$$F^+ \overset{\longleftarrow}{\underset{\longrightarrow}{}} F + H^+$$

$$Ka = [F] [H^+]/ [F^+]$$

$$\log Ka = \log [F]/ [F^+] + \log [H^+]$$

$$-\log Ka = \log [F^+]/[F] -\log [H^+]$$

$$(-\log Ka = PKa \; ; \; -\log [H^+] = pH)$$

$$Pka = \log [F^+]/ [F] + pH$$

$$\boxed{\log [F^+]/[F] = PKa - pH} \qquad (37)$$

Figure AppB.2 shows the relative % of each form of 3-malvidin glucoside at equilibrium in different pHs.

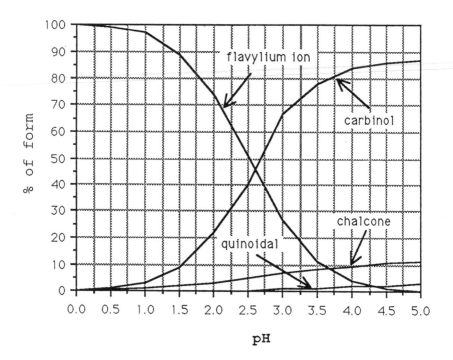

Figure AppB.2 - pH dependence of 3-malvidin glucoside forms at equilibrium (25°C).

It is clear that at wine pH range (3.0 - 3.8), the red color flavylium concentration, relative to its total pigments concentration is between (27 - 6)%.

By using Eq.37 and the pKa data from figure AppB.1 and figure AppB.2, let us check some of the equilibriums at different pHs.

- At pH = 1.6
  $\log = [F^+] / [F_j] = 2.6 - 1.6 = 1.0 ==> [F^+] / [F_j] = 10.$
  $[F^+]$ is the flavylium ion concentration.
  $[F_j]$ is the carbinol molecule concentration.
  At this pH the ratio of the red flavylium ion to the colorless carbinol (as shown in figure App B.2) is very high (about 9), which finds its expression by the very deep red color at such low pH.

- At pH = 2.6
  $\log = [F^+] / [F_j] = 2.6 - 2.6 = 0 ==> [F^+] = [F_j].$
  This can be roughly seen in figure AppB.2 where the flavylium ion (about 48%) is equal to the carbinol (42%) plus chalcone (6%). (the carbinol itself is in equilibrium with chalcone).

- At pH = 3.6
  This pH is in between the two Pka values (2.6 and 4.25).
  Let's check each of them differently :
  a. $\log [F^+] / [F_j] = 2.6 - 3.6 = -1.0 ===> [F^+] / [F_j] = 1/10.$
  It can be seen in figure AppB.2 that the flavylium ion is about 10% at this pH, where the carbinol (78%) and chalcone (9%) give the ratio of about 1/10 between the colored and colorless compounds.
  b. $\log [F^+] / [F_2] = 4.25 - 3.6 = 0.65 ===> [F^+] / [F_2] = 4.5.$
  $[F_2]$ is the quinoidal molecule concentration.
  Figure App B.2 shows that at this pH the flavylium ion is about 10%, and the quinoidal is about 2% which bring their ratio closely to about 5.

- At pH = 4.25

  Here too the two cases will be checked :

  a. $\log [F^+]/ [F_2] = 4.25 - 4.25 = 0 ===> [F^+] = [F_2]$,

  which can be seen in the figure that the concentration of both compounds is equal at about 2.5% .

  b. $\log [F^+]/ [F_1] = 2.6 - 4.25 = -1.65 ===> [F^+]/ [F_1] = 1/40$.

  At this pH the carbinol is about 85%, the chalcone 10% and the flavylium ion is about 2.5%, which gives an actual ratio of flavylium ion to the other two, to about 1/38. In this pH, only about 2.5% of the anthocyanin is red colored.

Appendix C - Wine Tasting Score Chart

The need to evaluate wine quality in the process of winemaking is more than just analyzing its components and figuring out how to improve it, or how to blend it.

Assuming the winemaker has done his best, and the wine is finished, the question now is how much is the wine worth in the marketplace, or in other words, at what price to sell the wine. If for example, the wine is a medium, well made wine, but its price is high, the knowledgeable customer will not buy the wine a second time if he perceives that the price is too high. On the other hand, if the wine is really excellent, and you underestimate it, you lose money by selling the wine at a low price. So, a comparative blind tasting can be a method to get some reliable answers to what is the value of the wine in the marketplace.

To fulfill this task, the evaluation method must be a quantitative one, not just qualitative or descriptive evaluation.

In such a test, one can buy 5-7 bottles of the same kind of wine under study (same variety and vintage), from different wineries, at all price ranges. By performing a blind tasting of all such wines (including the tested one), with a reliable and experience panel of judges, it is possible to find out the range of recommended price.

The most common tasting score is the Davis score of 20 points, or the modified Davis score (see table AppC.1).

DAVIS SCORE		MODIFIED DAVIS SCORE	
Parameter	Weight	Parameter	Weight
appearance	2	appearance	2
color	2	color	2
aroma, bouquet	4	aroma, bouquet	6
volatile acidity	2	astringency	1
total acidity	2	total acidity	2
sweetness	1	sweetness	1
body	1	body	1
flavor	2	flavor	2
bitterness	2	bitterness	1
general quality	2	general quality	2
total	20	total	20

Rating: 17 - 20 superior
13 - 17 standard
9 - 12 below standard
1 - 8 unacceptable

Table AppC.1 - "Davis" wine appreciation score chart.

Some people find the Davis score table unsuitable in these respects:

1. Too high weight for *color and appearance* (20% of the total).

2. Too low weight for *flavor* (only 10%).

3. *Sweetness*; It is not clear if it is a bad or good criteria, which in general depends on the type of wine.

4. No relation at all to *defects*. There may be some defects in the wine, which the score should relate to more specificaly (e.g. acetic, over oxidation, sweetness when not expected, off after-taste, sulfur-dioxide, hydrogen-sulfide, moldy smell and more).

5. *Body*; This parameter, like sweetness, depends on the type of wine and its style. In heavy Cabernet Sauvignon for example, it is

considered a good parameter to have full body wine, where in young fresh white wine or light red (Beaujolais style) it is of no meaning. This parameter may be better expressed in a general quality section if necessary, and be judged according to the wine type and expectations from such style.

6. The same argument holds for *astringency*, which depends on the wine type and its age. In white wines, for example, where astringency is very rare, it gets automatically one point (5%) just for the obvious reason of not being so (like sweetness in dry wine). Also, the lack of bitterness is not a reason to score wine for its quality. These parameters (astringency and bitterness) should be included in a defect section, or in the general quality section.

7. Too low weight for personal expression of the judge in the *general quality* section (only 10%).

Most of these difficulties are taken into account in a new tasting score table which I suggest here (see table). This score is based on 20 points of weight, where each parameter is scored in a range of zero to 5 points (to allow enough range of appreciation). Zero is the lowest score and 5 is the highest. Each parameter score (from 0 --> 5) is multiplied by its weight factor.

$$\boxed{\text{parameter value} = \text{parameter score x weight factor}}$$

(e.g. if one gave a score of 4 points for appearance and the weight factor of appearance is 2, the appearance value is 4 x 2 = 8).

The maximum total points are therefore, 20 x 5 = 100 points, an easy basis which is more convenient for many people to compare. In constructing the table, all the critic points mentioned above have been taken into consideration by choosing the parameters and their weight factors.

Quantitive Wine Appreciation Chart

Name :	No.					
	Type					
	Winery					
Date :	Vintage					
	Price					
Appearance (turbid,clear,brilliant)	*x2*					
Color (light,normal,dark)	*x1*					
Varietal aroma,Bouquet	*x6*					
Flavor	*x3*					
Total acidity (low,normal,high)	*x2*					
Defects (acetic,oxidized,moldy, sweet,oaky,sulfur-dioxide, sulfide,bad after-taste)	*x2*					
General quality (Body, astringency,age, balance,sever defect)	*x4*					
Total						

Score each parameter : 0 ➤ 5 ; then multiply by weight factor x

Rating :

95 - 100	superior
85 - 95	excellent
75 - 85	good
65 - 75	standard
45 - 65	below standard
0 - 45	unacceptable

remarks : -

- -

- -

- -

- -

Two parameters have to be explained in more details. One is the defects parameter, where minor defects can be related. If no defect is noticed, then the score is 5 points and the value for this parameter is 5 x 2 = 10. If any defect is noticed, then 0 to 5 range appreciation is available.

A very major defect which is beyond the x2 weight factor of *defects*, can be related in the next section - *general quality*.

In this section the judge can relate to any specific or general aspect such as after-taste, the wine's body, its astringency, bitterness, age, balance, severe defect, and any other personal impressions about the quality of that wine, good or bad. This section has x4 weight factor which arms the judge with a wide range of freedom.

After totalling the score, there are six groups of quality, and as our experience of using this score table shows, it uses almost the whole spectrum of scoring values, from very low numbers to almost 90 - 95, which makes differentiation between wines easier.

The chart also includes some identification data of the tasted wine. In blind tasting only numbers are available, and the type of wine. Other data (as winery, vintage, price) can be filled out after disclosure of the labels.

Appendix D - Cellar Recording Sheets

This set of cellar forms will help keep the wine's records from the harvest up to bottling. The recording data contains four sheets:

1. Must data, analysis, mechanical processing and additives.

2. Fermentation data.

3. Running analysis and wine processing data.

4. Bottling data (analysis, tasting score, stability tests and bottling details).

Receiving Record

Harvest

variety [   ]   vineyard [   ]
date [   ]   time [   ]
quantity [   ]   price (ton) [   ]

Must analysis

Brix [   ]

TA (g/L) [   ]

pH [   ]

Mechanical treatmen (mark v)

crushing [   ]   skin contact (hr) [   ]

draining [   ]   lees filtering [   ]

settling [   ]   pressing [   ]   heat excanging [   ]

Additives to must

SO$_2$ (ppm) [   ]   bentonite (g/L) [   ]

ascorbic (mg/L) [   ]   DAP (mg/L) [   ]

acids (g/L) [   ]   [   ]

Remarks : _____

_____

_____

_____

Fermentation Record

volume of must ⬚   tank No ⬚

Temp at starting ⬚

yeast strain ⬚   starter quantity ⬚

Date	Temp.	Brix	Remarks

Analysis After Fermentation

Date				
alcohol (%)				
TA (%)				
pH				
VA (mg/L)				
SO_2(ppm) free — total—				
residual sugar (%)				

Treatments (racking , transfering , fining , stabilizing , filtering , blending . additives , temperature changing , barrel aging , others)

Date	tank location	Treatment	Remarks

Bottling

A. analysis before bottling

Date [ ]

alcohol (%) [ ]   Residual sugar (%) [ ]

TA (%) [ ]   pH [ ]   VA (mg/L) [ ]

SO$_2$ (ppm)   → free [ ]   Dry extract (g/L) [ ]

→ total [ ]   [ ]

B. tasting remarks

Date [ ]   Ave. score [ ]

Panel [ ]

Remarks [ ]

C. stability test

Date [ ]   results [ ]

[ ]

D. bottling

Date	No. of cases	test sample remarks

Appendix E - Laboratory Equipment and Materials

To start and operate a laboratory in the winery for the basic analysis of must and wine, one should have the elementary equipment and chemical materials.

The following list of equipment, which is the expensive part of establishing the laboratory, was carefully selected, and is the minimum equipment needed to be able to perform the analysis described in chapter R.

As for the chemicals and calibrated solutions, we recommend to have them in stock at all times, especially toward harvest.

1. Equipment
   refractometer for field tests
   pH-meter + electrode
   microscope (ocular x10, objectives : x8, x40, x100)
   grid-slide for microscopic counting
   balance (precision 0.1 gram)
   analytical balance (precision 0.001 gram)
   ebulliometer
   hot plate + magnetic stirrer (controlled heat and speed)
   small laboratory centrifuge
   hand screw-crusher
   2 automatic burette (25 ml)
   hydrometers :
   a. set for Brix measurement
   b. for alcohol determination   0.980 --> 1.000
   c. for SO_2 solution   1.000 --> 1.050
   d. for dry extract   1.000 --> 1.020
   e. general use   0.900 --> 1.000
   thermometers : 0°C -->100°C; -5°C -->50°C; 0°C -->200°C
   distillation systems:
   a. for alcohol (500 ml)
   b. for ammonia (250 ml)
   c. cash distillation (500 ml)
   (for details see chapter R).

thermostated bath (range -5°C to 100°C)
pipettes: 1, 2, 5, 10, 20, 50, 100 ml (some of each)
erlenmeyers: 100, 150, 250, 500, 1000 ml (some of each)
beakers: 50, 100, 250, 500, 1000 ml (some of each)
one gallon beaker
volumetric flasks: 100, 250, 500, 1000 ml (some of each)
graduated cylinders:10, 25, 50, 100, 250, 500, 1000 ml (some of each)
funnels: 250, 500 ml.
teflon coated magnetic rods (different sizes)
picnometer 10 ml.
thermostated oven (up to 150°C)
refrigerator
laboratory stands, clamps, cleaning brushes
coffee maker

2. Materials and Chemicals
    sodium-hydroxide (NaOH)
    hydrochoric acid (HCl)
    sulfuric acid (H$_2$SO$_4$)
    hydrochloric acid 0.1N standard solution
    buffers (pH 4.00 and 7.00)
    indicators: phenolphtalein, methylene-blue, methylene-red,
        indigo-carmine, bromocresol-green.
    tartaric acid.
    potassium-hydrogen-tartarate (COOH-CHOH-CHOH-COOK)
    malic acid
    formic acid (HCOOH)
    t-butanol ((CH$_3$)$_3$-COH)
    citric acid
    potassium-chloride (KCl)
    iodine (I$_2$)
    potassium-iodide (KI)
    potassium-permanganate (KMnO$_4$)
    sodium-thiosulfate (Na$_2$S$_2$O$_3$)
    sodium-oxalate (Na$_2$(COO)$_2$)
    copper-sulfate (CuSO$_4$.5H$_2$O)
    boric acid (H$_3$BO$_3$)

trichloroacetic acid (CCl$_3$COOH)
magnesium-oxide (MgO)
sucrose
boiling stones
starch powder
sugar testing kit
activated carbon
microscope slides + cover glass
filter papers (different sizes)
capillary tubes
cheese cloth
distilled water

Appendix F - Bibliography

Most of the books written on the subject of "wine" deal with the history, the regions, tasting, consuming, touring, dining, education, general information, encyclopedia, social aspects, scandals and any angle of view on the fascinating subject. The actual list of books available on the market exceeds several hundred different titles.

On the other hand, very few books have been written in English on the technology of winemaking. Currently there are only about two dozen books on wine technology available.

The books we have found to be most useful are:

- *Table Wine* - M.A. Amerine, M.A. Joslyn; Berkeley, 1960, 1970.
- *General Viticulture* - A.J. Winkler et al; Berkeley, 1962, 1974.
- *Dessert, Appetizer and Related Flavored Wines* - M.A. Amerine, M.A. Joslyn; Berkeley, 1964.
- *The Technology of Wine Making* - M.A. Amerine et al; Connecticut, 1966, 1971.
- *Applied Wine Chemistry and Technology* - A. Massel; London 1969.
- *Phenolic Substances in Grapes and Wines* - V.L. Singleton, P. Esau; New York, 1969.
- *Chemistry of Winemaking* - Editor A.D. Webb; Washington 1974.
- *Wines, Their Sensory Evaluation* - M.A. Amerine, E.B. Roessler; San Francisco, 1975.
- *Grape Growing* - R.J. Weaver; N.Y. 1976.
- *Analysis of Must and Wine* - M.A. Amerine, C.S. Ough; New York, 1979.
- *Commercial Winemaking* - R.P. Vine; Connecticut 1981.
- *Wine Production Technology in the U.S.* - Editor M.A. Amerine; Washington, 1981.
- *Grape Growing and Winemaking* - Davis Jackson, Danny Schuster; New Zealand, 1981.

- *Biotechnology:* Textbook of Industrial Microbiology - W. Crueger and A. Crueger. Translated from German,1982.
- *Principles of Fermentation Technology* - P.F. Stanburg, A. Whitaker; N.Y. 1984.
- *Knowing and Making Wine* - Emile Peynaud. Translated from French, New York, 1984.
- *The Taste of Wine* - Emile Peynaud, Translated from French, San Francisco 1989
- *Modern Winemaking* - Philip Jackisch; Ithaca and London, 1985.
- *Vines, Grapes and Wines* - Jancis Robinson; New York, 1986.
- *Practical Microscopic Evaluation of Wines and Fruit Juice* - H.R. Luthi. Translated from German, 1981.